THE GATEKEEPERS OF SYCAMORE SPRINGS

Kansas Historical Journey to "Healing Water"

Bonnie Dornes Hanni

Anamcara Press LLC
Lawrence, Kansas

Published in 2023 by Anamcara Press LLC
Author © 2023 by Bonnie Dornes Hanni
Cover and Book design by Maureen Carroll
Palitino Linotype, Minion Pro, Corbel, DINosour
Printed in the United States of America.

Book description: Embark on a journey through the generations in *The Gatekeepers of Sycamore Springs,* where history and healing waters converge. The legacy and evolution of a public resort is revealed through the personal stories of the author.

All rights reserved. No part of this publication may be reproduced, distributed, or transmitted in any form or by any means, including photocopying, recording, or other electronic or mechanical methods, without the prior written permission of the publisher, except in the case of brief quotations embodied in critical reviews and certain other noncommercial uses permitted by copyright law. For permission requests, write to the publisher, addressed "Attention: Permissions Coordinator," at the address below.

ANAMCARA PRESS LLC
P.O. Box 442072, Lawrence, KS 66044
https://anamcara-press.com

Ordering Information:
Quantity sales. Special discounts are available on quantity purchases by corporations, associations, and others. For details, contact the publisher at the address above. Orders by U.S. trade bookstores and wholesalers. Please contact Ingram Distribution.

ISBN-13: 978-1-960462-34-3 The Gatekeepers of Sycamore Springs, (Paperback)
ISBN-13: 978-1-960462-35-0 The Gatekeepers of Sycamore Springs, (Hardcover)

BIO023000 BIOGRAPHY & AUTOBIOGRAPHY / Adventurers & Explorers
BIO028000 BIOGRAPHY & AUTOBIOGRAPHY / Cultural, Ethnic & Regional / Indigenous
BIO022000 BIOGRAPHY & AUTOBIOGRAPHY / Women
Library of Congress Control Number: 2023951568

THE GATEKEEPERS
OF SYCAMORE SPRINGS

DEDICATION

I dedicate this book to my parents, Charles and Mary Dornes, who gave me a full, happy life living at Sycamore Springs. And to my brother, Dan, who shared this playground with me. We have many shared memories.

I dedicate the effort to writing this book to my faith in Jesus Christ. After 9 years of research and collecting information, I was able to start writing this book.

He saw me through every step of the way. Every day, I would say out loud: Okay, Lord, where do I go with the story today? Through this practice of waiting on direction, I was able to write this book in four months. Grace was given to me through this faith.

— Bonnie Dornes Hanni

CONTENTS

FOREWORD xi
1. Alice Gray Williams Story 1
2. Stories of the First Kansas Indians 13
3. Stories of the Kansas American Indian 21
4. Kansas Nebraska Homesteads 29
5. Westward Expansion 35
6. Civil War 43
7. Lawrence, Kansas—Battleground 43
8. Underground Railway, Nemaha County, Kansas 53
9. James Lane 63
10. Sycamore Springs 71
11. 20th Century 85
12. 1935 - 2017 97
13. Our World in turmoil 107
14. Today 119
15. Sycamore Springs Community 139
16. 1984 - 2017 155
17. New Generation 161
References 169
Photo Citations 175
Author Biography 183

TRIBUTE

Langdon Livengood was a historian at heart. He shared so much information with me about his personal stories of Morrill, Kansas, Sycamore Springs, and the surrounding community. This inspired me to continue writing this book.

Unfortunately, Langdon passed away before the book was published but I extend my tribute to his family.

Langdon and Elda Livengood

FOREWORD

This is a book about Sycamore Springs and "the healing waters." A progression of documented events brings us to the present day. It is a collection of observations along the Kansas historical journey that lead us to the sycamore land.

History gives us perspective, but the future is still untold. People, places, and time determine the past. Our ancestors set the course. What and where and how they lived brings every one of us to the present.

Alice Gray had a story too. It has made an impact on many lives. Her history has given her a perspective of exploration and determination to understand the healing waters. Her story remains the beginning of my story.

This well-worn path is part of my history. My family is associated with Sycamore Springs in a very small way. We owned and lived at this historical resort from 1949-1966. We made it our home. We added to the vast history of Sycamore Springs. Everyone who came to Sycamore Springs contributed to this story with your memories; I am so thankful for each of you. You, the reader, are a vital part of our history by coming to enjoy Sycamore Springs while we lived there.

The book *The Gatekeepers of Sycamore Springs* is not only a history book but adds a human perspective through my eyes of how we lived in this place. I hope you see the personal story of Sycamore Springs with humor and joy, and it makes you think of a memory or two also.

This story begins many years ago around the 1860s, but my story starts in 1949. See you there.

Chapter 1
Alice Gray Williams Story

**Alice Gray Williams,
1860-1934**

Life in Kansas with the Kickapoo Tribe

The Journey to Sycamore Springs begins here. Alice Mabel Gray was born in October 1860 to John and Anna Gray of Hiawatha, Kansas. She grew up in the area and received her education in Hiawatha, Kansas. She was the only daughter of 6 children.

She was raised among the Indian tribes that roamed the state of Kansas in those days. Alice became friends with the Kickapoo Indian Tribe that lived nearby. The Kickapoo Indian Chief named Chawkeekee valued the Gray family friendships.

Chief Chawkeekee of the Kickapoo tribe took a personal interest in little Alice. Because of her sunny disposition and her red hair (the Indians attached superstitious importance to red hair), Alice soon became a great favorite among the friendly Indians.

Messages sent by Chief Chawkeekee were arranged to visit John Gray at his home.

Chief Chawkeekee had mentioned mysteriously the great Indian secret about "The Healing Waters."

Pony Creek, Sycamore Springs

Sycamore Tree, Sycamore Springs

Ruins of Lawrence, Lawrence Western Town

Parents of Alice Gray Williams

NEAR HIAWATHA, KANSAS, IN BROWN COUNTY, Alice Mabel Gray was born in October 1860 to John Wesley Gray and his wife, Anna Marie McKeon (McCune, name discrepancy as recorded in the "History of Nemaha County" by Ralph Tennel, 1916).

Historical Background in the Life of John Gray and his wife, Anna Marie McCune (McKeon):

John Wesley Gray was born December 7, 1826, to Anson Gray (1797-1870) and Jane Harris (1807- 1847) in Seneca County, Ohio.

John Gray was married in Illinois in 1857 to Anna Maria McCune (McKeon).

Anna Marie McCune (McKeon), born in New York in 1837 was left an orphan at the age of twelve years of age, and then she made her home with a cousin. The cousin was a newspaper editor who came to Lawrence, Kansas in 1854 to help edit the "Herald of Freedom" newspaper. Anna Marie McCune came to Lawrence with her cousin.

Anna Marie was living in Lawrence, Kansas when the town was sacked and burned by the Quantrill Raid pro-slavery ruffians on August 21, 1863, and she lost all earthly belongings. She then went to Illinois with a pro-slavery family named McVeigh and was later married to John Gray.

Six sons and a daughter were born to the marriage of John and Anna Marie (McKeon) McCune Gray: Alice Mabel Gray (1860-1934); Anson Elsworth Gray (1862-1928), Los Angeles, California; Arthur W. Gray (1865-1953), Medford, Oklahoma; Walter (no record), Grant, Oklahoma; Delbert Eugene Gray (1874-1944), Muscotah, Kansas; and Frederick Thurston Gray (1876-1960), Florence. Kansas The Gray children were prosperous and were known for being upright worthy citizens of their respective communities.

Alice's father, John Wesley Gray, was one of the earliest pioneers in Kansas, coming from New York in 1857 at the age of

31 years. There was only one house in Hiawatha at the time the Grays came to settle in the area. They located a homestead one mile north of Hiawatha and lived there for many years.

At the outbreak of the Civil War, John Gray joined a company of guards organized in Brown County, Kansas for service in the Union Army under Capt. I. J. Lacock. The Union company journeyed to Atchison, Kansas, and tried to be mustered in as a part of the First Kansas infantry but was disappointed in not being accepted and turned away. After three months of being away from their homes, the company disbanded.

John Gray then enrolled in the militia but was rejected at Leavenworth, Kansas. James Pope was captain of the company in which he enrolled. Still desirous and anxious to serve his country, he joined the home guards, under Lieutenant Perkins, and assisted in repelling the Price Invasion of Kansas. In later years he was always proud to relate the fact that he took part in the expedition which resulted in General Price and his rebel army being driven from Kansas.

Large Sycamore Tree, Sycamore Springs

While Gray was away in the Union service, his wife and mother of his children, Anna Marie (McCune) Gray cribbed 1,000 bushels of corn, and cut and hauled the winter's fuel from the woods seven miles away. The family lived all alone and was unprotected from the Indians while John was away.

John possessed a roving disposition. He was one of the original "Forty-Niners" and crossed the plains of Kansas to the gold fields of California during the great gold rush of 1849. He returned home via Cape Horn. He went on many freighting expeditions to Pike's Peak, Colorado, and was known as an old Indian fighter.

In 1905, John Gray died in Oklahoma. Mrs. Annie Maria McCune (McKoen) Gray died in 1886. Both lie buried in the Fairview Cemetery, Goff, Kansas.

Life at the "Healing Water"

ALICE GRAY AND HER FATHER JOHN HEARD many tales about the springs and were greatly curious to see them. After much persuasion, Chief Chawkeekee consented to take them to the springs within a period of five "sleeps" (five days).

True to his promise, on the fifth day, Chief Chawkeekee appeared at the door of John Gray homestead, riding one pony and leading two others. Alice and John mounted the ponies, and they were off on their journey to the spring of "healing waters."

After winding around hills and trees following a faint Indian trail, they arrived in sight of the end of their quest. A strange sight met their eyes. Along the banks of a beautiful winding creek (Pony Creek), running full of sparkling clear water, breaking into pools of water and ripples over the rocks, were many Indian wigwams (lodges). The Indians and their children were moving about their daily life. Above this temporary encampment arose the brow of a steep hill, formed of peculiar blue clay. Under the foot of this hill, the waters of two great springs poured forth and fell in a cascade into the creek.

Kickapoo Sweat Lodge

Blue Clay Mud, Sycamore Springs

Soon the little party passed through the encampment and stopped in front of the springs. Chief Chawkeekee invited Alice and her father to "drink and grow strong." They did so and found the water to be pleasant-tasting mineral water, crystal clear.

While Alice and her father were there, they watched the Indian people using the water. The people drank a gallon or more a day while others took "sweat baths" with the water.

The sweat wigwam (lodge) was an air-tight wigwam into which the Indian people rolled heated stones and then poured water over them. This gave them the effect of the modern steam bath.

George and Alice Gray Williams

They would also roll up in the soft, blue, clay mud and let this remarkable substance "draw" the poisons out. The Indian people attributed the amazing results they were getting to supernatural powers, not realizing that the day would come when modern science would verify their discoveries in present-day laboratories.

As Alice later said: "Many is the time that I have seen people brought to these springs on stretchers, supposedly hopeless and beyond human aid, and seen them go away in a short time able to walk and no longer afflicted with Rheumatism or its kindred ailments. I have all of my life wanted every sufferer in the country to know the benefits to be expected from a stay at these amaz-

George Williams, 1916

ing natural health springs. I have felt that if the time ever came when the word could be spread to sufferers throughout the land, I would feel that my own life had not been in vain, bringing to the attention of everyone the wonderful value of Sycamore Mineral Springs." (Interview with Ralph Tennel, 1916.)

Alice Williams lived to see her dream come true with radiant good health and happiness that comes with good health. She said, "You step into the stream, (taking part in the present moment) but the water has moved on." (Life continues to move on.)

Alice and her father returned home from this visit and the impression of that day never left them. As she grew older, this association with the Indians grew continually closer until she was at times permitted to sit in some of their tribal council, a privilege accorded few white men and almost no white women.

Alice Gray & George W. Williams
Family Life

UPON REACHING YOUNG WOMANHOOD, ALICE Gray married George William Williams and shared married life to the end of their days.

George W. Williams, a farmer in Oneida, Kansas, Nemaha County, was born in 1847, in Moniteau County, Missouri, and is a son of Eli W. and Eliza (English) Williams, natives of Pennsylvania and Moniteau County, Missouri respectively. Eli Williams was a son of James Williams of Pennsylvania, and was born near the city of Harrisburg, Pennsylvania. The Williams family emigrated from Moniteau County, Missouri to Kansas in 1855, and made a settlement on a farm of 160 acres near Deer Creek, in Nemaha County, known by the name of "Williamsdale."

The Williams' farm was one of the oldest farms in Kansas and has been owned by members of the Williams family for many years. Here on the unbroken prairie lands, the settlement was one of the first pioneer settlers of Nemaha County.

Eli took a prominent part in the early struggles in Kansas and was here during the beginning of territorial difficulties. He was elected a member of the Kansas legislature scheduled to meet at Lecompton, Kansas. Richard Clency was his bodyguard and a trip to Lecompton had been planned, and it was arranged that the two men go on horseback to the meeting. They had good horses saddled, with saddle bags and canteens for water. A sack was thrown across the back of the saddles, which contained flour, bacon, and a frying pan, gun, and hatchet to complete the outfit for the trip.

On the morning of their proposed departure, James Lane (a Jayhawker) sent messengers to them, telling them not to start, as they would be killed, and to defend themselves as best they could until he (Lane) could meet them.

It was a time of trial and trouble for the family. George and his sister, Fanny, stood guard all night at the cabin door with an axe and knife handy, ready to trade their lives in defense of their deceased mother, Mrs. Eliza Jane Williams Johnson.

The parents and members of this noted pioneer family were all members of the Christian Church.

George Williams was only seventeen years old when his father died, and he was left to help an invalid mother rear a family of boys and girls in a new and barren country. With ox teams, he helped break up the virgin sod. With four yoke of oxen, he hauled all the family supplies from the Missouri River. He also hauled lumber with ox teams to build the first drug store erected in Seneca, for Dr. McKay.

The Overland Trail to the Far West passed through Oneida and Seneca at this time. The great wagon trains of gold seekers were constantly passing through on their way to the mountains of California. Many of the Pony Express riders and the old United States Rangers were well known to him.

George had often seen large herds of deer on the land where Oneida now stands. He had many times seen hostile bands of Indians decked out with paint and war regalia. They were looking for trouble, but no depredations were committed by the savages any nearer to the Williams' home than the Little Blue River.

Pilgrim Fathers boarding the Mayflower

Betsy Ross Flag, American Revolution

Williams developed the homestead into a fine productive farm and prospered during the many years in which he had been a resident of Kansas. In knowing the number of years of residence, he was probably the oldest living pioneer citizen of Nemaha County at that time. He was a stockholder in the Farmers' Shipping Association of Oneida and a charter member of the Knights and Ladies of Security.

George traveled over a considerable portion of the United States with his wife, Alice, who accompanied him part of the time while she was engaged in the United States Indian Service. He Married on November 23, 1881, to Alice Mabel Gray, a pioneer teacher of Brown County, Kansas, the union was blessed with a son Laurin and a daughter Maude, who died in infancy.

Their son Laurin L. Williams was born in Oneida in 1883 and resided on the historic Williams farm. He served for five years as a rural mail-free delivery carrier out of Seneca but liking the farm life as more suited to his tastes, he returned to his home place.

He was educated in the common schools of Kansas and finished a course in painting at Campbell University, Holton, Kansas. He had some talent as an artist who loved the outdoor life and was a lover of all animal life. He kept a kennel of thoroughbred dogs and his pack of wolfhounds as his pride and joy.

Laurin Williams
1883-1967

GEORGE SERVED AS AN INDUSTRIAL TEACHER at Tuba, Arizona in the Navajo country and held that position with the Kickapoo

Indian tribe at Horton, Kansas. He was a government farmer at Tuba, for a time and did a great amount of good on behalf of the Indians because he was practical in his instructive work. He taught the Indians from what knowledge he had accumulated from many years of experience tilling the Western soil.

Alice taught school during the greater part of her mature life, and in later years she was engaged in Indian service. Her first appointment in Indian service was at Tuba, Arizona as a teacher among the Western Navajos. She was lovingly called the "Soniskee" by that mighty tribe of 24,000 members. At her request, she transferred to the Great Nemaha School of Iowa Indians and worked for several years on the reservation in Brown County, Kansas, not far from her early home.

When the Oneida, Kansas post office was placed under the civil service, Mrs. Alice Williams took the examination held to select someone to fill the position, and she was appointed postmistress of Oneida. Her husband, George, was her able assistant in taking care of the duties of the post office.

Alice was a stockholder of the Best Slate Company of Mena, Arkansas, owned land near White Cloud, Kansas, and had western property and a nice residence property in Seneca, Kansas. She was a member of the Knights and Ladies of Security of Seneca, Kansas and a long-time working member of the Woman's Christian Temperance Union, serving as president of the Seneca Union for three years. She was also a member of the Women's Relief Corps of Seneca for some time and was a patriotic instructor of the corps.

At that same time, she was the camp guardian of the Camp Fire Girls of the Minnehaha Camp at Oneida. The collection of Indian relics possessed by Mrs. Williams was the finest in this section of the State, and she took great pride and infinite pains in adding to her noted collection for many years.

Speaking from an ancestral standpoint, the paternal grandfather of Mrs. Alice Williams was Anson Gray, a native of New York, and a direct descendant from the Grays of old Revolutionary stock. Her maternal grandmother was Jane Harris of Pennsylvania, whose father was John Harris, the famous founder of Harrisburg, Pennsylvania and a direct descendant of an ancestor who came to America with the Mayflower contingent. Members of the Harris family fought in the American Revolution on behalf of independence.

Chapter 2
Stories of the First Kansas Indians

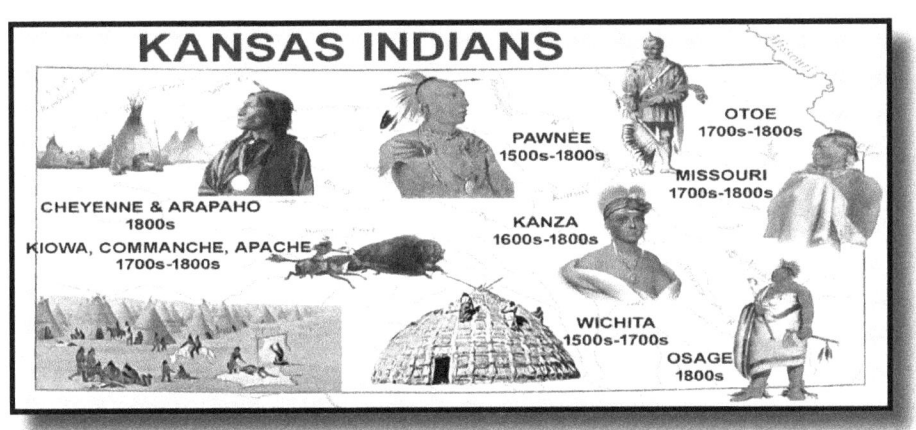

Native Americans of Kansas/Kansas Indians Map

There are many stories passed down through the William's family, and much history about encounters with Native people of the region. The family took pride in their positive relations with people of the Sac & Fox, and other Tribes.

There has been much written about the American Indians. No nation has had as much written about them as the American Indians. They were picturesque in their daily costumes. Oral traditions and storytelling contribute to this history. Folklore and legends are common ways of communicating history. Native communities possess rules and procedures for their storytelling. Their dress was beautiful, and their handiwork very primitive, yet so grand that it is part of the way they communicate their story.

They are a foundational race to this country, and their memory will be forever perpetuated in the names which have been given to our towns, counties, states, mountains, rivers, and lakes. Though we have never had a reservation located in Nemaha County since becoming a county, Oneida, Nemaha County, Kansas sound sweet to us, and it is all because these tribes lived or moved through this Indian country.

The Indians in Nemaha County provoked neighbors in their "way of life," not understanding how the pioneers created life on the prairie differently from them. No one has ever told of trouble from them with, but for one or two brilliant exceptions. An occasional connection with Indian troubles came to Sabethans, however. Joseph Prentice, a Sabetha farmer, unearthed a treasure, resulting from an Indian raid of early times.

Indian trouble occurred in the nearby Nebraska territories when Prentice was an early-day merchant. During land trades, he came into possession of a Nebraska farm where the raid occurred. A story was known for years about the Indian attack on a party of emigrants on the way to fortune in the far West. A man named Wilcox buried a can of money on the farm. His brother searched the ground over and over for the money after the death of the man who was wounded in the Indian fight. The farm, as a farming proposition, had not been considered of much worth during this time. But one day Joe Prentice determined to get something out of his trade. He took to deep excavation when he plowed his ground. On a rather steep incline near the house, he plowed up a rusty apple or tomato can. It was found to have $2,136.50 in silver and gold. Joseph Prentice said that the real lesson in this was that "any farmer will turn up money if he plows deep."

Nemaha County related to a real Indian tragedy, although our own Nemaha County Indians did not commit the crimes. In August 1874, it was the Cheyenne who attacked Nemaha County travelers when they were traveling to Colorado over land.

John German with his five daughters and one son, were in Chautauqua County traveling in their covered wagon to

The Gatekeepers of Sycamore Springs | 15

Colorado hoping to acquire benefits for the health of their daughter, Catharine.

Catharine and the brother were driving the cows some distance at the rear of the wagon. As they came over a hill, they saw the wagon attacked by the Indians. The father, mother, and one sister were killed before the horrified eyes of the boy and girl. The Indians saw them and killed the boy. Four sisters, Julia, Sophia, Adelaide, and Catharine were then carried by the Indians for four days, with but one stop for food. The story makes me think of the movie Dances With Wolves with the people on the prairie who met up with the Indians."[1] *Bonnie Hanni quote*

During their travels, they passed a soldier's camp. Once two of the girls were left behind with two Indians and when the soldiers overtook the main band, Adelaide and Julia were not with them. The older sisters thought they had been killed. The Indians had simply abandoned them on the prairie to starve.

Adelaide and Julia wandered over the prairie until they came to the soldiers' camp where they found an old blanket, corn, and crackers, and for six weeks the little children lived on these abandoned scraps. They found hackberries that grew plentifully and drank the clear spring water at hand. One night they awoke to find themselves covered with leaves. Doubtless, some animal, already satisfied as to appetite, covered the little girls for future use.

Finally, the little girls were discovered by soldiers and were so dirty that the men wouldn't believe they were white girls. The men wept when the young girls told of their sufferings. (Later, when the little round-eyed girls attended school in Sabetha, their playmates hung on every word of this experience as they told it again and again.)

Meanwhile, Catharine and Sophia had been separated, the former accompanying the Cheyenne into New Mexico and Sophia going to Colorado with a band of Arapaho. By the time Catharine reached the Texas border, she had lost track of time, and hope of being recovered. But when she met Chief Stonecalf in Texas, her hope revived, for the great chief was grieved at the attack on her people. "I will try to take you home to your people," he said, "but it will take long, long." And he did return them.

Not long after this meeting they began to move eastward, and it did take "long, long." It was winter and there was snow on the ground. Many braves died of hunger. One night when they reached a canyon with good water and plenty of wood, Indians from other bands came straggling in, and with them to Catherine's happiness, came Sophia.

In some way, Sophia had heard of the rescue of the little sisters, and that General Miles was searching for the two older ones. Although the girls were not allowed to be together, they were kept in the same camp. And a few days later, Chief Stonecalf told them that the Indians had decided to tell of the killings and gave themselves up to the white chief and were to take Adelaide and Julia back.

When they reached General Miles' camp, the Indians were lined up and the girls pointed out which ones were in the original band who killed their parents, brother, and sister. These Indians were sent to St. Augustine, Florida. General Miles took the guardianship of the girls for two years when they were taken first to Lawrence and later to Leavenworth.

Dog Chief, Pawnee Indian Scout

In Leavenworth, Pat Corney became the girls' guardian, and moved the sisters to Granada Township in Nemaha County, and later to Sabetha where the younger girls graduated from Sabetha schools. Catherine married Amos Swerdfeger, a brother of Mrs. Corney and later. recalled, "When we reached the soldiers' camp all the soldiers were lined up and cheered us. I still feel a lump in my throat when I think of it. I thought I had never seen such white people; they looked as white as snow. Since we were with the red people for some time that was why they looked so white and pretty."

All four girls married and settled in different areas. Catherine and her husband moved to California. Julia became Mrs. Brook, also of California; Mrs. Frank Andrews (Adelaide) lived in Berwick, Nemaha County; and Mrs. Albert Feldman (Sophia), near the Nemaha County line in Richardson County, Nebraska.

There were no resident Indians in Nemaha County. The Kickapoos adjacent on one side of the county are in Brown County, and the Sac and Fox tribes have always been in Jackson County. The twenty miles on either side of the Nemaha River may have been exempted from Indian claims, resulting in the Indians never taking up a residence in Nemaha County. The Nemaha River runs north and south near the center of the county. Nemaha County is forty miles wide. But the Indians have always made frequent and, invariably friendly, calls on their white Nemaha County neighbors. The latest call happened within a short amount of time and is an interesting illustration of the Indians' acceptance of modern conditions and their endurance of primitive traditions at the same time.

Lucette Goslin, the little six-year-old daughter of native Indians, Mr. and Mrs. Johnny Goslin of the Indian res-

Sabetha Saint Anthony Hospital

ervation located in Jackson County, was brought to the Sabetha hospital by her mother. The child had been swinging while she had in her mouth a small wheel from a toy train of cars. The wheel became lodged in her windpipe, and she was taken to the hospital for its removal, under the modern, advanced surgical conditions and surroundings suitable for her surest recovery.

Meantime, Mrs. Goslin, the mother, took a room at a hotel. During the night she gave birth to a baby. The next morning, she got up, wrapped the new little papoose in approved Indian fashion, visited her little daughter at the hospital, and returned to her reservation with the new member of her family. But the little girl remained at the hospital a week longer to recover from her throat trouble. Trust was built between the mother and the hospital staff.

The Indians in northeastern Kansas were generally peaceful and friendly. Many years ago, a son of Tohe, an Iowa chief whose Sac & Fox reservation was still at White Cloud in Doniphan County died with honors, and many white friends attended to mourn with the Indian brothers the loss of "a good Indian." A custom of placing the body in a sitting posture on the surface of the ground upon the top of a high hill, with his face to the setting sun and bows and arrows, a war club, and a pipe near him to cheer and protect him on the Long Journey. His pony was shot and placed beside him. They were covered over with a mound of earth, a white flag raised, and charms placed around the mound. Doniphan County is filled with such mounds and is a real mine for Indian collectors. But not one such Indian mound is known to exist in Nemaha County.

It was an annual event for Nemaha County people to press the self-starter of their automobiles and spin over to the Kickapoo Indian reservations for the pow-wow. Each year the pow-wow became more and more like an American event. The best baseball games of northeastern Kansas are now played on these occasions by select Indian college boys who attend school at Carlisle or Haskell (Lawrence, Kansas) and whose parents live on the reservation. The women's

"Squaw ball" and Indian boy ball games, however, remain very interesting events, and old and young Indians from six years to sixty enter both games.

But as late as 1884 it was more than a couple of hours to drive to the reservation. C. H. Isely of Spring Grove tells of a trip made to the reservation from his farm near Sabetha in August of that date. The driving part of the way was even roads across the open prairie, and through unfenced lands which now were worth from $100 to $200 an acre.

The care and conduct of the Indians were criticized by Isely at that time, a condition which vastly improved, except for the fact that the worst road in northeastern Kansas runs through the Government lands on the reservation. It is said it is the only section of the State without a road drag implement. The farms of the Indians themselves, however, were well kept in 1884 and continued to be well maintained.

These are just some of the stories passed down through the William's family about interactions with the first Kansas Indians.

Indian Village 1879

Chiefs on Horseback

Chapter 3
Stories of the Kansas American Indian

Kanza Men

The conversation narrated by Alice Gray Williams to Ralph Tennel, Sr., 1916:

Alice, whom the Indians delight to call, "Soniskee," meaning "Our Good Red Mother."

The old Indian tribes had no written history. Their history was passed from father to son. From some of the oldest Indians now living I have gained knowledge of Indian traditions, customs, and life.

It is said by these Indians, and history bears them out in their statements, that the first Indians of Kansas were a part of the Great Dakotah Tribe, and that they came here with the great bands of Indians who migrated from north of the Great Lakes. They wandered around for many years and finally settled on the Missouri river and its tributaries.

They were called the Kanzas or Kaw Indians and the Osages. The Kanzas had as their territory the land from Nebraska on the north to Arkansas on the south and all west of the Missouri river. The Osages were to have Missouri and all the land along Missouri and along the Osage River, and part of their hunting grounds extended into Kansas.

For many years they dwelt in this manner, but they were unfriendly. Fair maidens were stolen from tribe to tribe, as they were not allowed a peaceful marriage, and this alone caused endless trouble. They spoke the same tongue, and their tribal affairs were managed in the same manner.

In 1806 our government helped them to make a peace treaty with each other which each tribe kept sacred, and then they combined forces against the hated Pawnees and the Whites, who were intruding on their hunting grounds. Their depredations became so numerous and so serious that the Government called a Council near the present site of Atchison on an Island called Ise Au Vache (French word meaning "Cow Island"), or Buffalo Island. This council was a great affair. It is said that there were some 150 Kanzan and 13 Osages there, representing their powerful and mighty tribes. Officers of the garrison were present. The Council was closed.

Peace prevailed and the peace pipe filled with Kln-ni-ki-nick was smoked and the Indians kept their promises, and no depredations were ever committed by them. These Indians believed in the Great Spirit or Waconda, and they believed in life beyond the grave. They were honorable in their family life and were kind to their squaws and children. Let me say right here, an Indian never strikes his child. No whipping is allowed in their homes or schools. The women managed the household affairs and did the work, but be it said in the old time Indian life the squaws did the "bossing" around the wigwams but had no voice in the affairs about the warpath, or to the lands, or their tribes.

Postcard: Greetings

The first treaty between the United States and these tribes was made in 1815. In this treaty, the past was blotted out and forgiven and these tribes recognized our government and pledged their loyalty to it.

In 1825 the United States Government made a treaty with them for the cession of their lands in Kansas and Missouri. In this treaty they ceded all the lands in eastern Kansas: "Beginning where the Kansas River empties into Missouri to the northwest corner of Missouri, thence to the Nodaway River, thirty miles from its entrance into the Missouri River; from there to the entrance of the Nemaha River into Missouri to its source, which took in the present county of Nemaha. From here to the source of the Kansas River, then on to the ridge dividing the Kansas River from Arkansas, and on to the west border of Missouri and with that line thirty miles to the place of beginning."

The United States agreed to pay them $3,500 per year for twenty years, either in money or merchandise. Additionally, they were to furnish the cattle and hogs and farm implements, a farmer, and a blacksmith. Thirty-six sections of land on the Big Blue were to be sold and the money from that sale was to be kept for the use of their schools.

In 1846 the Kanzas and their neighboring tribes ceded all their lands to the United States Government.

From this time on they began to deteriorate. They were moved to Oklahoma and the climate did not agree with them there. I am told by the oldest Indians now living that there are now, but a few poverty-stricken ones left, of this once wealthy and powerful tribe, from which the fair State of Kansas derived its beautiful Indian name. Kansas in the Indian tongue means "Smoky."

At this time the Pottawatomie Indians had no home in the United States. They gave them the land of the Kanzas for their homes. It contained 576,000 acres.

The Pottawatomie Indians were in possession when our forefathers came here. They were peaceful Indians, and their lands were the hunting and playgrounds for the mighty southwestern tribes. Buffalo and deer were plentiful, and

Indian Chiefs

the prairie was covered with rich grass.

These tribes were what was known as the "Horse Indians" because they had ponies. Many tribes had no horses at that time.

"Chama' ___ meaning "grandma" in the Indian tongue, told me her mother said that a day's ride west from the Missouri river, there were, once some Ground Indians, who lived in holes dug deep down and that they covered them over with poles and skins and that when these Indians left or were driven west that the covering dropped in, and so made the holes we call buffalo wallows.

Aunt Lizza Roubidoux Barrada, a great-granddaughter of Joseph Roubidoux, the founder of St. Joseph, Missouri says that when she was a girl and when Chama was a girl that the Pawnees came here a day's ride to the west of her home at the mouth of the Great Nemaha, and stayed and lived for several years, and fought the Iowas. She says the Iowas whipped them so completely, that they went away and never came to fight the Iowas again.

A Pawnee burial ground is still pointed out to the visitor on the Iowa Reservation, on the Great Nemaha River. Skulls and arrowheads are found there to this day. Chama says that was said the number of Pawnees was like the leaves upon the trees. The Pottawatomies were allotted and some of them took land of their own and some went to Oklahoma. Some went to a reservation in Jackson County, Kansas, where many of them still reside. George Williams, who is one of the oldest settlers of this vicinity, says when he was a small

Kickapoo Sculpture, Horton, Kansas

lad many Indian tribes passed through Nemaha County to visit other tribes.

Hundreds at a time could be seen winding along the trails, along the creeks. Sometimes there would be a bunch go into camp and hunt and fish and then, like the Arabs of old, would "Silently fold their tents and steal away." They were silent people. Sometimes they would sing and dance their war dances to amuse the boys and girls who would call upon them. The Indian is a very matter-of-fact person and does not often joke, yet sometimes he will play a little joke.

The End of the Interview.

"The games played by them on the ground where Oneida now stands were Indian ball for the boys and squaw ball for the women and girls. They measured their strength with these games, with each tribe always trying to be the winner. An Indian cherishes his ball bat as well as his gun or bow and arrow, and he always takes it with him on any visit he makes to other tribes.

"But the old Indian has passed away and only the young progressive Indian is to be found here now. They are quietly living on their reservations."

Kickapoo Nation History

PART OF THE KICKAPOO NATION HAD MANY years of migration before settling in Northeast Kansas. Before contact with Europeans, the Kickapoo lived in the Great Lakes region. Beginning in the 1640s, the Algonquin tribes (which Kickapoo was a part of) in this region came under attack from the eastern tribes. By 1658 the Kickapoo had been forced west into southwest Wisconsin. In 1770 the tribe moved to the Illinois area, near Peoria. After the wars with the

Buffalo, Kickapoo Nation Reservation, Horton, Kansas

Americans and the settlement of the Ohio Valley, they signed treaties in 1819, signing over their remaining land east of the Mississippi River and relocating to southern Missouri (1819-1824).

In 1832 the Missouri Kickapoo exchanged their reserve for lands in northeast Kansas. They were skilled farmers and used hunting and gathering to supplement their basic diet of corn, squash, and beans. Relocating first to Missouri and then Kansas, Prairie Bands of Kickapoo scattered across the plains warning other tribes that the white man was coming to Kansas.

The Kickapoo felt the change coming to their area. They had been forced from their homelands for many years into the lands in which they were required to relocate their groups. Though the treaties temporarily brought peace and set aside land specifically for these tribes, the wave of American settlers slowly but surely infringed upon native space once again. Many other tribes were feeling the change coming to them too. The people were guided by the "will- to-survive." With the European influx, along came smallpox epidemics. The Kickapoo were experiencing this devastation within their tribe, causing their numbers to decrease.

The development of the western states devastated the rights of the Native American communities. Prospecting for gold brought many people through this territory on their way to Colorado and California. Having seen thousands of pioneers traveling through, their stock of buffalo depleting, and the availability of the grasses and cutting down the stands of river timber, they were convinced that the pioneers and freighters were irresponsible. They became uneasy. The plains for the tribe were the winter resting places because of the milder winters, and it became much harder to stay with the settlement of the pioneers in their area.

The Kickapoo were evicted from their lands piece by piece. Many of them moved to Mexico. In 1834 the army was able to move the last groups of Kickapoo out of many of the areas. One day the Americans would occupy most of the land.

With the passage of the Kansas-Nebraska Act in 1854, white settlements poured into Kansas. The Kansas Kickapoo signed a treaty in May 1854 selling their excess lands (600,000 acres) for $300,000. They also agreed to accept either allotment or relocation to the Indian Territory (Oklahoma). This decision was unpopular with many Kickapoo, and in 1857 another group left for Mexico, picking up dissatisfied Potawatomi members and Seminole along the way. By this time, the number of Mexican Kickapoo had grown to more than 1,000, while there were only 300 Kickapoo in Kansas.

The Kickapoo eventually left the area of Sycamore Springs. The wagon train travelers coming through took over the area as a stopping point for the natural spring water making it uneasy for the Kickapoo to stay.

"The legend that I learned as a little child was that the little Indian children played with the pioneer children and gave away the secret of the healing water. Eventually, the pioneers took over the springs pushing the Indians to move from the spring area."

The Kansas Kickapoo fought hard to maintain their tribal government and managed to delay the implementation of allotment until 1908. Since then, they have only managed to keep 19,200 acres.

The Kansas Kickapoo still live on the lands from the 1832 treaty. The tribal government office is located in Horton, Kansas, and organized under the Indian Reorganization Act (1934) that was approved in 1937. In 1951, they barely managed to avoid a government attempt to terminate their tribal status.

Currently, there are four recognized bands of the original tribe first encountered in the Great Lakes: The Kickapoo Tribe of Indians of the Kickapoo Reservation in Kansas, the Kickapoo Tribe of Oklahoma, the Traditional Kickapoo Tribe of Texas, and the band of Mexican Kickapoo still in Coahuila Mexico.

The Kickapoo Tribe of Indians of the Kickapoo Reservation in Kansas is a federally recognized tribe of Kickapoo people.

Chapter 4
Kansas Nebraska Homesteads

Sod House Homestead

Homesteads & Pre-emptions
Kansas & Nebraska (1872)

I. What is a Homestead?

It is an uncultivated farm given by the United States to any man or woman who lives on and cultivates it for five years. The whole cost is $18.00 for 160 acres, and $4.00 of this is not payable for five years. Outside of the Railroad Limits, the size of a Homestead, free to anyone, is 160 acres. Within the Limits, it is 80 acres, except for Soldiers who have served 90 days, who are allowed to take 160 acres in the Railroad Limits.

II. Who may become a Homesteader?

Any man, or any woman that is any native of legal age, and any foreigner who has declared his intention to become a citizen, which any immigrant may do on the very day that he lands in America.

III. How does one become a Homesteader?

He goes to the United States Land Office and has free access to maps showing all the vacant land in the region. Having chosen the land he thinks will suit, he goes and examines it, returns to the Land Office, makes an application for it on a form furnished to him by the officer there, pay the fees for recording, (at most $14.00), and is then master of the land he has chosen. All this business he can do — though not as well — through the Clerk of the County in which the land lies. The Homesteader must begin to occupy his land within six months after his application is put on record, and he may journey away from his land at will, if not absent more than half a year at once and provided that he fixes his residence nowhere else.

IV. Can a man become the full owner of his farm sooner than at the end of five years?

Yes. After six months of residence, he can at any time purchase his land by paying the Government price, the highest of which is $2.50 per acre, and the lowest half of that sum.

This is called PRE-EMPTION HOMESTEADS. They now abound in Nebraska, and the building of railroads caused thousands to pour in there during 1871, and the crowd will be larger in 1872. Kansas also has lands still open to Homesteaders, particularly on the Line of the Atchison, Topeka & Santa Fe Railroad. (Ralph Tennel, Sr. History of Nemaha County Kansas 1916-Sod House)

Wagon Train

The availability of cheap land and the desire to own one's farm was a powerful inducement to Easterners to move west. Emigration to the Western territories became so rapid that Congress, beginning in 1830, passed preemption laws which allowed squatters to settle a claim before it went through the public land sale system.

In 1841, the Preemption Act was passed which allowed settlers to stake a claim of 160 acres and to purchase it after 14 months of settlement for one dollar and twenty-five cents an acre. Promotional materials flooded the market and stimulated emigration with glowing reports of fertile land, a healthy climate, and cheap land. The information appeared in guidebooks, as advertisements in pamphlets and newspapers, and in paper advertisements and circulars. In addition to boasting about the West, these types of documents also provided useful information about the public land sale system.

The role of the United States government in the sale of its public lands cannot be overstated as a factor in westward migration. Millions of acres came into the public domain as Loyalist lands that were confiscated after the Declaration of Independence and as states ceded their land claims to Congress during the ratification of the Articles of Confederation. These vast tracts of land were a valuable asset to a new country deeply in debt.

Beginning in 1784 the Continental Congress passed a series of ordinances that provided for the surveying and sale of its newly acquired Western territories. The government further encouraged the emigration of individual settlers through the availability of easy credit and generous terms. Veterans of the Revolutionary War, the War of 1812, and the Mexican War were also allowed to purchase land cheaply. By 1819, frontier land could be purchased as noted earlier for as little as one dollar and twenty-five cents an acre.

C. Albert White, A History of the Rectangular Survey System (Washington, D. C.: Government Printing Office, 1983), 9-16.
Deborah Kmetz, U.S. General Land Office Surveyor's Field Notes.

Edmund Needham Morrill
Kansas 13th Governor 1895-1897

AT ABOUT THIS TIME IN THE 1860S, CONGRESSMAN EDMUND Needham Morrill endeavored to get a bill through Congress removing the Kickapoos from Brown County to Wisconsin. It failed. Occasionally the matter was brought up for discussion, but nothing was done. The Indians were peaceable, well-behaved neighbors, as industrious as many of their white friends, and people generally saw no reason why they should be taken from the home of their fathers and placed elsewhere.

The pensioning of soldiers of the State militia was first introduced from this district through Congressman Edmund Morrill. Morrill asked to have three soldiers pensioned who lost their legs through freezing conditions when called out by Governor Osborne to stop an Indian uprising in the southwestern part of the State in 1873. He finally secured fifty dollars a month for the three men, establishing a precedent that it was the regular soldier's duty to enter such fights and that if State soldiers were injured, they should be rewarded.

Edmond Needham Morrill

A one-term Republican (13th) governor of Kansas, 1895-1897, Edmund N. Morrill was born in Westbrook, Cumberland County, Maine on February 12, 1834, and attended school in his hometown, graduating from Westbrook Seminary in 1855. Morrill then worked as superintendent of the Westbrook schools in 1856 and 1857 before moving to Kansas Territory in 1857. He settled in Brown County where he erected a sawmill, served as a free-state member of the territorial legislature in 1857 and 1858, and enlisted on October 5, 1861 as a private in Company C, Seventh Regiment, Kansas Volunteer Cavalry — the "notorious" Charles R. Jennison and Daniel R. Anthony commanding. Morrill almost immediately was promoted to sergeant (on October 10, 1861), and within a year (on August

27, 1862) to captain and commissary of living resources.

After mustering out of the service as a major in October 1865, Morrill served as clerk of the district court of Brown County (1866-1870) and county clerk (1866-1873), and in 1871 he founded the county's first bank, serving as its president from 1887 until his death.

Excerpt from Speech given at the State of Kansas legislature 1895:

> "In conclusion, I would say that I have unbounded confidence in the future of Kansas. As it has steadily advanced in the years that have passed, always true to its motto, constantly pressing forward and upward, I believe it will continue to advance. It has vast hidden resources that will be developed in the future to add untold wealth to its present means; millions of acres of fertile soil which have yet been untouched by the plow, inviting the honest yeoman from other states. With proper encouragement, sturdy men can be induced to make their homes within its borders, and countless millions can be secured to develop its boundless resources. Then let every citizen of the state, without regard to his political affiliations, be loyal to its interests, and ever ready to defend its fair name." E. N. Morrill

(Governor biographies courtesy of the Kansas Historical Society.)

Lewis and Clark Expedition

The Missouri Compromise of 1820

Chapter 5
Westward Expansion

Thomas Jefferson & Napoleon

Louisiana Purchase 1803

In 1803, President Thomas Jefferson purchased the territory of Louisiana from the French government for $15 million. The Louisiana Purchase stretched from the Mississippi River to the Rocky Mountains and from Canada to New Orleans, doubling the United States' size. To provide enough land to sustain this independent ideal population of virtuous yeomen, the United States would have to continue to expand. The United States' westward expansion is one of the defining themes of 19th-Century American history.

By 1840, nearly 7 million Americans—40 percent of the nation's population—lived in the trans-Appalachian West. Following a trail blazed by Lewis and Clark, most of these people left their homes in the East in search of economic opportunity.

Like Thomas Jefferson, many of these pioneers associated westward migration, land ownership, and farming with freedom. In 1843, one thousand pioneers took to the Oregon Trail as part of the "Great Emigration."

President Jefferson commissioned the Lewis and Clark expedition in 1810 as a "literary pursuit." Jefferson concealed the expansion from England, France, and Spain. He was afraid he might lose the claims and rich resources that the West offered.

Lewis and Clark

Lewis and Clark began their expedition in St. Louis, Missouri.

The Missouri Compromise Slavery

MEANWHILE, THE QUESTION OF WHETHER SLAVERY would be allowed in the new Western states shadowed every conversation about the frontier. In 1820, the Missouri Compromise attempted to resolve this question. It had admitted Missouri to the Union as a slave state and Maine (wanting to separate from Massachusetts) as a free state, preserving the fragile balance of states in Congress. More importantly, it stipulated that in the future, slavery would be prohibited north of the southern boundary of Missouri (the 36°30' parallel) for the rest of the Louisiana Purchase.

The Missouri Compromise of 1820

HOWEVER, THE MISSOURI COMPROMISE DID NOT apply to new territories that were not part of the Louisiana Purchase. The issue of slavery continued to fester as the nation expanded. The southern economy grew increasingly dependent on "King Cotton" and the system of forced labor that sustained it.

Meanwhile, more and more Northerners came to believe that the expansion of slavery impinged their liberty of land ownership. This included citizens, the pro-slavery majority in Congress who did not seem to represent their interests, and yeoman farmers. They did not necessarily object to slavery itself, but they resented the way its expansion seemed to interfere with their opportunity of developing economic success.

The Kansas-Nebraska Act
Bleeding Kansas

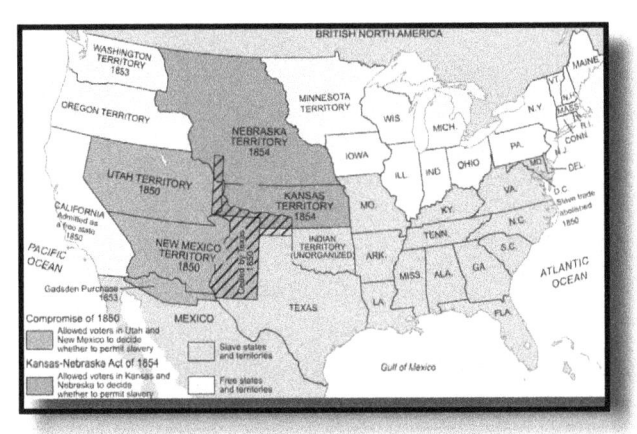

Kansas-Nebraska Act Map

IN THE POLITICAL ARENA, ARGUMENTS BETWEEN the Democratic Party which supported popular sovereignty and states' rights, and their opposition, the Whigs, heated up and had lasting effects leading up to the outbreak of the Civil War. Animosity grew between the North and the South.

For decades, both northern states and southern states had threatened secession and dissolution of the Union over the question of where slavery was to be permitted. At issue was power. New states were organized into self-governing territories before they became states. Each state developed a position on the slavery issue before its admission to the Union. Kansas was the first territory in which they could decide to accept or reject the slavery issue.

Voters had a choice between a constitution that allowed slavery and maintained the increase of the institution, or a constitution that banned the importation of new slaves but allowed for generational increases in the number of slaves. Eventually, the voters in Kansas rejected the Lecompton Constitution and an irreparable split in the Democratic Party developed. The non-slavery parties reorganized and became the Republican Party.

Constitution Hall, Lecompton Ks First Kansas Territorial Capital, Lecompton Ks

Lecompton, Kansas was near the city of Lawrence and was the first territorial Capital of Kansas. Lecompton was founded in 1854 on a 640-acre Wyandotte Indian land claim on the south bank of the Kansas River. The town, which was originally named Bald Eagle because of the many eagles that nested along the river, was renamed later that year to honor Judge Samuel D. Lecompte, the chief justice of the Kansas Territorial Supreme Court. In 1855, the territorial legislature chose Lecompton to be the only official and permanent capital of the Kansas Territory, however, Topeka, Kansas later became the official capital of the state.

Constitution Hall, Lecompton, KS

First Kansas Territorial Capital, Lecompton KS

Border Ruffians vs. Free-Staters

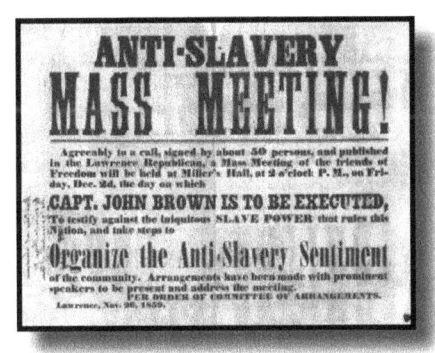

Anti-Slavery Mass Meeting Poster 1859

ON THE GROUND, THE FIGHTING OF THE "border ruffians" in "Bloody Kansas," such as John Brown's raid on Pottawattamie Creek, violently settled what each side's Washington Congress counterparts were debating- slave or free. The turmoil in Kansas continued when President James Buchanan appointed a new territorial governor and asked him to have the citizens adopt a constitution and apply for state admission.

While many Free Soilers lived in Kansas by this time, most boycotted the constitutional convention, held in the temporary capital in Lecompton, Kansas. The Free Soilers' historic slogan called for "free soil, free speech, free labor, and free men. This attracted small farmers, debtors, village merchants, and household and mill workers who resented the prospect of black labor competition—whether slave or free—to the Lawrence, Kansas 1859 territories.

Stephen Douglas Democratic Politician

STEPHEN A. DOUGLAS (1813-1861) WAS A U.S. politician, leader of the Democratic Party, and orator who espoused the cause of popular sovereignty about the issue of slavery in the territories, including Kansas before the American Civil War (1861-1865).

He was re-elected senator from Illinois in 1858 after a series of eloquent debates with the Republican candidate, Abraham Lincoln, who defeated him in the presidential race two years later. Douglas developed a strong interest in the western territories. Some of his first legislative proposals were territorial expansion, the construction of a Pacific railroad, a free land (homestead) policy, and organizational territorial governments.

Throughout his tenure, Douglas continued to argue for the doctrine of popular sovereignty—the right of the people of a state or territory to decide the slavery question for themselves as a Union-saving tactic. Douglas's popularity waned as the party system floundered on the slavery question.

Stephen Douglas,

He was proposed as the Democratic presidential candidate in 1852 and 1856, but did

not receive the nomination until 1860, when it was too late, and as noted, Abraham Lincoln became the 16th President of the United States.

Abraham Lincoln Arrives in Kansas Territory

ABRAHAM LINCOLN CAME TO LEAVENWORTH, KANSAS, the largest Kansas town, on December 3, 1859 at the request of his cousin, Mark Delahay's wife, Louisiana. Mark Delahay had confidence that the visit would have a dramatic effect on the election to decide slavery or free state because of Lincoln's stand against slavery.

At the time, there were too many people for slavery, so the impact of his speech was not well received. Presidential hopeful Democrat William Seward came to Kansas in 1860 and was better received. The battle for freedom was not easy for Kansas.

The United States Congress was afraid of the Kansas slavery issue and what it would do to the nation. It was not a favorite subject in 1860 to allow Kansas into the Union.

Kansas, situated on the American Great Plains, became the 34th state on January 29, 1861. Its path to statehood was long and bloody.

Diligence and Dedication—Life of Abraham Lincoln

- "It is amazing what Abraham Lincoln did and from where he came."
- "The road to success was not an easy one for Lincoln as said in The Great Emancipator's Triumphs and Tragedies.
- Lincoln was born into abject poverty in a one-room log cabin on Feb. 12, 1809, in the waning days of the Thomas Jefferson administration. Yet he managed to rise to power, becoming the 16th president of the United States.
- Lincoln had no formal education. In fact, he even dropped out of grade school. His father, Thomas Lincoln, wanted him to become a farmer and frontiersman. Lincoln refused because he disliked the hard labor associated with frontier life.
- Mostly self-taught, Lincoln was an avid reader, having read and reread, most notably, the Bible, the works of William Shakespeare, and Aesop's Fables. A self-educated lawyer, Lincoln eventually earned his law license in 1839 and went into private practice in Springfield, IL.
- At the age of 23, Lincoln bought a general store in New Salem, IL in 1832. The

Abraham Lincoln

business wasn't successful, and he went bankrupt; it took years for him to pay off his debts. It was good for history that he did not prosper as a shopkeeper; this failure pushed him on toward other goals.
- He lost his first love, Ann Rutledge when she died in 1835 of typhoid fever. Lincoln suffered what is sometimes called a nervous breakdown. In fact, he suffered from depressive tendencies throughout his life.
- Lincoln ran for the U.S. Senate and lost twice. He also ran for the U.S. House of Representatives and lost twice before finally getting elected in 1846 as President of the United States.
- The failures deepened his resolve to make life better.

Seal of Kansas

Abraham Lincoln with Flags

Chapter 6
Civil War
Lawrence, Kansas—Battleground

Massachusetts Street, Lawrence, Kansas

"The Civil War came early to Missouri and Kansas, stayed late, and was always characterized by unremitting and unparalleled brutality. More than anywhere else, it was truly a civil war." —Bo Kerrihand, America's Civil War

Missouri counties that bordered Kansas were strongly pro-slavery and wanted their neighbor, Kansas, to be a slave state too. Over 1700 men from Missouri came to Kansas to vote (fraudulently) for pro-slavery. Missouri pro-slavery guerrillas including bushwhackers fought to retain the right to keep slaves.

Bushwhackers were characterized by their lawless methods to support the cause including ambush, plunder, arson, and murder. The Missouri Ruffians and bushwhackers crossed

Main Street on Fire, Lawrence, KS

into Kansas and raided Free State communities to further their cause by shooting, burning, and hanging those opposed to slavery. Lawrence, Kansas was one of the main communities that were raided.

Bushwhackers

THE BUSHWHACKERS WERE MISSOURIANS WHO fled to the rugged backcountry and forests to live in hiding and resist the Union occupation of the border counties. These guerrilla fighters harassed, robbed, and sometimes murdered loyal Unionist farmers on both sides of the state line.

Men joined with the bushwhackers for several reasons. Some, like Frank James, had been paroled from the Missouri State Guard, and upon returning home they were constantly harassed by the "jayhawker" troops garrisoned in the border counties. Other ruffians, like his younger brother Jesse, sought safety in the brush at a young age and thus grew into the tumultuous and violent life of a

Bushwackers

warrior bandit. Still others, like William Quantrill, were landless drifters, "border ruffians," or bandits from the border war who sought personal gain from the complete chaos of the Civil War.

Quantrill's Raid

WILLIAM CLARKE QUANTRILL WAS A Bushwhacker who formed his group of pro-Confederate guerrillas. He had been a gambler as well as a schoolteacher in Lawrence before siding with the Confederates. Quantrill became the most feared and famous of the guerrillas who killed Union troops. It worked both ways with the brutality. Missouri and Kansas ran red with blood. The buildup of these atrocities resulted in the massacres at Lawrence, Kansas on August 21, 1863.

The Raid in Lawrence, Kansas between the Union Free-Staters and Missouri farmers began 6 weeks after the Civil War battle in Gettysburg. Lawrence had been settled for 9 years. Settlers had come from Indiana, Ohio, and Illinois to settle in Lawrence.

William Quantrill

One of the interesting connections between Nemaha County with the border war is the fact that a prominent citizen, H.C. Haines of Nemaha County came from Canal Dover, Ohio, which was the boyhood home of William Quantrill. H.C. Haines, of Sabetha, said that Quantrill was a boy who had no "folks." He came out West with a family by the name of Beach. Beach was living near Lawrence. No one seems to know what became of Quantrill, but he possibly moved to Texas.

Quantrill's Raiders included future outlaws Frank and Jesse James and Cole Younger. His group fought and killed Union soldiers and Free Staters along the Kansas and Missouri border and elsewhere.

John Brown was a controversial abolitionist who also arrived in Kansas territory with a commitment to end slavery. Quantrill was opposed to John Brown's earlier activity.

The culmination of Quantrill's violent raid, mentioned earlier, was what became known as the Lawrence Massacre. In 1863, Quantrill led approximately 450 Confederate raiders into Lawrence, a Unionist and antislavery stronghold that was also home to Republican Senator James Lane who had become a target of pro-slavery forces. He was the leader of the Kansas Jayhawkers. Although Lane escaped, Quantrill's men killed many in cold blood before looting and burning much of the town. They cleaned out the banks and took all the whiskey from the taverns.

James Lane-Jayhawkers

Sabetha Connection to Quantrill

H.C. Haines had a general store in Sabetha, Kansas. It was the place to go for the latest fashions or any other fabric goods available.

> As a kid, I went with my Mother, Mary Dornes, to buy the latest fashion dress for Easter Sunday or special occasions. We went to Sabetha every Saturday night to do our shopping. Haines Store was a store that had many different types of articles (dry goods). I remember that when you paid for your item, there were overhead wires with mechanical containers with the money enclosed that were sent to the Office upstairs. Any change was then sent back down through this small box on a wire to the clerk. It was a very large store with updated fashions available. I also remem-

ber the store had soft carpeting on the steps and floor leading up to the clothing department. We would play on the 2 steps and would often hide in the clothing cubicles. The clothes were on pullout racks. They did clothing alterations and repairs for any purchase. Very classy store for Sabetha. (Bonnie Dornes Hanni, Submitted!)

Haines began business in Sabetha for himself in 1878. In 1880 he built a 26 by 80-foot store and developed a large steadily increasing business. The first floor was devoted to a full line of staple and fancy groceries and a partial line of dry goods and gents' clothing. Carpeting, trunks, and books occupied the second floor. A stock of from $20,000 to $30,000 was carried, and four or five clerks were employed.

Haines Clothing Store, Sabetha, KS, 1910

Jayhawk or Jayhawker

The origin of the Jayhawker remains unclear but was used during the unrest on the issue of slavery. It was later used solely as a label for free-state proponents and eventually became associated with all Kansans.

According to tradition today we associate the Kansas Jayhawk with the University of Kansas. When the University of Kansas fielded its first football team in 1890, the team was called the Jayhawkers. Over time, the name was gradually supplanted by its shorter variant, and KU's

KU Flag Banner

John Brown Mural

sports teams are now almost exclusively known as the Jayhawks.

I received my first degree from KU in 2002 as a non-traditional student (older than normal student age) in Fine Arts-Textile Design. Studied abroad in Kyoto Japan. Wonderful experience. I also received a Ronald McNair scholarship from KU for being the first-generation student in my family to attend college. I went to Colorado State University for my master's degree. Became a Professor at Illinois Institute in Chicago. (Bonnie Dornes Hanni) Rock Chalk, Jayhawk, KU!

John Brown
Abolitionist Against Slavery

TRAGIC PRELUDE BY JOHN STEUART CURRY illustrates John Brown and the clash of forces in Bleeding Kansas. (A mural painted by Kansan John Steuart Curry for the Kansas State Capitol building in Topeka, Kansas. It is located on the east side of the second-floor rotunda.)

John Brown was known as a hero in the North and a radical in the South. He felt he was the instrument chosen by God to abolish slavery. He believed he was predestined to bring an end to slavery, a sin against God. He was fighting a "holy war".

Congress opened the territories of Nebraska and Kansas (1854) with the possibility of them becoming slave states. Pro-slavery advocates infiltrated Kansas from Missouri.

Slavery was becoming a principal issue in Kansas. The Fugitive Slave Law of 1850 which required citizens to cooperate with the capture of fugitive slaves angered Brown and other abolitionists. In 1854, Congress enacted a law—the Kansas—Nebraska Act, that would further enrage more Northerners.

John Brown

John Brown and his five sons were all willing to die for the cause of abolishing slavery. Brown and his sons drove a wagon loaded with Beecher rifles he collected in Ohio and Illinois and arrived in Kansas in October 1855.

In 1858, Brown left Kansas to pursue support for the Southern invasion. He met a former slave, Harriet Tubman, who had led several dozen slaves north from Maryland. She shared

John Brown's anti-slavery beliefs. Brown expected Tubman to help recruit young men for his army and to encourage local blacks to join Brown when the time came to side with him.

Neither Harriet Tubman nor Frederick Douglass engaged in the raid. Tubman felt it would destroy the Underground Railroad and expose its methods, its routes, and all who participated in this endeavor. Douglass felt the venture would fail.

Having secured financial support from wealthy abolitionists, Brown returned to Kansas in mid-1858. On October 16, 1859, John Brown, from a Northern state rode into the industrial town of Harpers Ferry, Virginia located 61 miles northwest of Washington, D.C. at the junction of the Potomac and Shenandoah rivers. Virginia was considered a southern state. The plan was to seize the United States Federal arson of weapons located there. There was a musket factory and rifle works, an arsenal, several large mills, and an important railroad junction. The guns would be valued at 7 million dollars in the present day.

Mayhew Cabin, Nebraska City, NE

Brown was the first white man to enter a southern state in the name of abolishing slavery. His friend, Frederick Douglass, the African American orator, thought him to be a hard man. Brown had empathy for the slaves and that stood him apart from any other white person in the historical record.

Some of the men with John Brown were his 5 sons; farm boys; veterans (Free-Staters) of the guerrilla war in Kansas; an Oberlin College African American student; a pair of Quaker brothers from Iowa who abandoned their beliefs of pacifism; a former slave from Virginia; and men from New York, Connecticut, Pennsylvania, and Indiana. They were ready to make war on abolishing slavery.

Harpers Ferry transformed the nation. America was traumatized by the fear of massive slave rebellions felt by the Southerners, and the issue of pro-slavery confronting the Northerners who hoped that the confrontational issue of slavery would be delayed. Both sides believed this would end with a compromise but that never developed.

He brought together several slaves along with his sons, and the other men who came with him to help overcome the guards in Harpers Ferry. John Brown was going to equip the slaves with guns to lead a rebellion that would lead to the abolishment of slavery.

When Brown and his men attacked the guards, they were able to overtake

Mayhew Cabin Room, Nebraska City, NE

them but most of the slaves did not arrive to help. They were afraid of what might happen to themselves.

Soon nearly a thousand uniformed men emerged in the little town of Harpers Ferry and surrounded Brown's tiny band of men. President James Buchanan dispatched a company of 90 Marines from Washington, under the command of one of the Army's most promising officers, Lt. Col. Robert E. Lee, who was also a slave owner.

Mayhew Cabin Ladder, Nebraska City, NE

All five sons of John Brown were killed, and John negotiated with Lt. Col. Robert E. Lee in hopes of releasing the hostages and retreating across the river to Maryland. There was no compromise and the attack continued. Brown was severely injured but survived. The course of history would have changed at that moment if John Brown, the martyr, would have been killed.

The trial was all in the name of slavery. Brown was convicted of treason, first-degree murder, and promoting Negroes to produce a revolt. The verdict came back guilty with the death penalty.

John Brown, along with his other "patriots," was hanged on December 2, 1859, for his multiple deeds. Among the witnesses were Robert E. Lee, Thomas J. Jackson (who later became known as Stonewall Jackson), and John Wilkes Booth (who killed President Lincoln), a fanatical believer in "Southern patriotism."

The incident at Harpers Ferry spurred many militia groups to be formed in the South for fear there would be more slave uprisings. The South wanted to preserve the chains of slavery. The trained soldiers

John Wilkes Booth

were the first recruits for the Confederate Army in the Civil War (1861) between the North and the South.

Mayhew Cabin was built in 1855 at John Brown Cave, Nebraska City, Nebraska.

The Underground Railroad route traveled through northeast Kansas and southeast Nebraska towns such as Falls City and Nemaha County. Once escaping slaves and freedom seekers reached Nebraska City, they were hidden in the Mayhew Cabin. They crossed the Missouri River into the free state of Iowa en route to Civil Bend and Tabor, Iowa.

Chapter 7
Underground Railway, Nemaha County, Kansas

Slavery Wanted poster, 1860

Albany, Kansas

John Brown and James Lane used Albany as a waystation for hidden slaves.

The trip from Topeka to Iowa on the James Lane Trail took about 3 weeks. James Lane Freedom Trail is marked on Highway 75 north of Topeka. "I have even seen Lane Chimneys, (rock cairns) along the fence lines."

For those hiding slaves, it was a dangerous endeavor. The Fugitive Slave Act of 1850 made aiding fugitive slaves a federal crime punishable by six months in prison and a $1000 fine. Five years later, the pro-slavery Kansas territorial government enacted legislation saying any person who spoke, wrote, or printed materials to assist escaped slaves would be found guilty of a felony and sentenced to death. Additionally, those who helped slaves escape their masters would be committing grand larceny and face death or imprisonment with hard labor. Pro-slavery spies and slave hunters kept tabs on those suspected of aiding slaves. Therefore, it took dedication and conviction to hide and help slaves on their way to freedom.

Albany Stores

The 13th Amendment of the US Constitution abolished slavery in 1865. There are over 80 sites in Kansas: homes, churches, forts, cemeteries, and museums used in the Underground Railroad.

There was a house/hotel in Albany that hid slaves as they were passing through Kansas for freedom. In the house, a hidden trap floor compartment is where the slaves were kept in times of uncertainty.

In 1857 a colony of a dozen or more families related by blood and affinity, from Painted Post and Castle Creek, New York, settled at the head of Pony Creek, naming their town Albany in honor of the capital of their native state, Albany was two miles north of present Sabetha on the east edge of Nemaha County. Among these pioneers were the families of William and Samuel Slosson, John and William Graham, Noble H. Rising, John Tyler, George Lyons, Edwin Miller, and Elihu Whittenhall. The men were destined to exert considerable influence on the civil, military, and economic affairs of the two counties in the next quarter century.

Albany School

Educated, cultured, and possessed of sound business acumen, they were whole-hearted supporters of Free State principles.

Albany

WILLIAM B. SLOSSON WAS ONE OF THE ORIGINAL settlers in Albany. He was from Maine, Broom County, New York. He had an interest in protecting the slaves. His journey to Kansas was rigorous. He took the train to "the West" to Illinois. He left Illinois in 1857 for St. Louis,

Missouri where he boarded the steamer boat on the Missouri River with 520 passengers. It took 5 days on the river to arrive at Doniphan Kansas.

Traveling over the land in a spring wagon to Hiawatha took 3 days, with 8 more days to Morrill. He met E.N. Morrill and was given information to claims of 160 acres at $1.25 per acre. As William Slosson traveled on to Albany, he was carrying a Beecher rifle, an axe, a spade, a grip of clothes walking with a compass to a mark "Pony Creek" on the map.

The Slossons and Grahams quickly realized the potential of the Lane Trail and were instrumental in organizing a branch of the Underground Railway known as League No. 40. Many slaves escaped to freedom in the next five years, but no written records exist of names or numbers.

Some slaves reaching Albany in 1862 were provided homes and employment in the area. Among these were five Holden siblings; their mother; and two Russell siblings, Daniel and Lena Holden. Fanny Whittenhall, the wife of W. G.

William Slosson

Sargent, taught Jane Holden to read and write and maintained an extended correspondence with her friend long after Jane married a man named Scott and moved away from Albany. Charles Holden married Lena Russell, Cora Holden married Thomas Frame, and another sister married ex-slave John Masterson. One of the Holden brothers served in the Union Army and was killed in action. His mother eventually received $1800 in back pay and pension.

Although most of the freedmen moved elsewhere after the War, E. J. Holden owned a 10-acre tract on the east edge of Albany some 50 years later according to the 1912 Atlas of Nemaha County.

In February 1859, following the "Battle of the Spurs" (using spurs as their weapons), John Brown passed through Albany. He spent his last night in Kansas in the Elihu Whittenhall cabin, which he shared with family members and the only cabinet grand piano in Kansas Territory.

Elihu Whittenhall

On this journey, he hid a large group of slaves in the "White Banks" cave while hiding their horses and ponies in the nearby creek bottom. "Pony Creek" received its name as a result. The cave is located across the lake north of Sabetha on the east side wall. Timber covered the area making it ideal for hiding the travelers.

The following day William Graham escorted Brown's party to the Missouri River in Nebraska Territory.

A village of bygone days. Every year there is a celebration of "Albany Days," Experience every old-style working trade during these days. The older generations are interested in teaching the younger generation to keep these memories alive.

Nemaha County Grain Harvest

Sabetha, Kansas-Nemaha County

GOLD WAS DISCOVERED IN 1848 IN CALIFORNIA at John Sutter's fort near Sacramento. In 1858 gold was discovered near Denver, Colorado. By 1859 at least 30,000 persons traveled the Oregon Trail.

The Pony Express traveled on the Oregon Trail from 1860-1861. They went from St. Joseph, Missouri to Sacramento, California, 1,966 miles. The riders changed horses every 10 miles, exchanging mail, and riding for 75-100 miles at 10 miles an hour. It took 13 days to get from St. Joseph to Sacramento. (City of Marysville, KS is noted for the Pony Express trail.)

Pony Express lasted only 18 months due to the building of the Transcontinental Telegraph in July 1861. In 1863 the Union Pacific Railroad began construction from Omaha, Nebraska and the Central Pacific Railroad started construction from Sacramento, meeting in 1869 at Promontory Summit north of Salt Lake City, Utah. This was the beginning of the end of the Oregon Trail.

Sabetha was named by a very pious Biblical student who started across the plains to California to seek gold. His oxen died near the area on a Sunday. He performed the sad last rites over the grave of the oxen, naming the spot Sabetha as a euphonious substitute for the Hebrew word, "Sabbaton," which signifies Sunday.

The city of Sabetha was located south of Albany. This was a very rich agricultural region. It was said that there is health in the air and wealth in the products of the ground.

Sabetha was in a good location for people from neighboring communities to come. The _____ railroad came through the

Grain Harvest, Albany, KS

town. It was only 7 miles from Sycamore Springs.

There were three doctors: Dr. Cecil Hunnicutt, Dr. Virgil Brown, and Dr. Tom Montgomery. Dr. Martin Rucker was the eye, ear, and throat doctor at the clinic on Main Street in Sabetha. The Catholic Nuns managed St. Anthony Hospital.

Thrasher Tractor, Albany, KS

The Duckwall's Store had all the latest comic books, paper doll kits, candy served by the scoop (my dad loved peanut clusters), the latest hankies, sundries, teen magazines, ladies' magazines, and toys that we all wanted to spend our allowance on. Fifty cents for Danny and twenty-five cents for me per week.

Thrashing Machine, Albany, KS

Mr. Harold Geiger, owner of Geiger Jewelry and Dave Duey, owner of Duey Jewelry, were the stores for jewelry and decorative items. White Variety, LaDuskies, and Haines Fashion Center were the places to buy your clothes. Men's clothing was purchased at George Hughes Clothing. The Weiss Shoe Store owned by Elise and Frieda sold shoes; and Red Emert Shoe Center carried all the latest shoe designs.

Greene's Drug, Miller Fountain Drug, and Darville Drug Store carried all the prescriptions that were needed. We bought our books for school at Greene's Drug. You could even get weighed on the big scale there. The drug stores had different flavors of dipped ice cream and soda pop served at the counter. According to Harla Havemann Estele, having a Green River or a Peanut Hickey, or chocolate or cherry Coke for 10 cents was one of her favorites. This was the place to hang out to see all the boys. We stayed as long as we could by sipping our coke through the straw, one straw slurp at a time wait-

Thrashing Days, Albany, KS

Main Street, Sabetha, KS
1930 - 1940

ing for the good-looking guys to come in.

Safeway Grocery Store, HyKlas Grocery with Wally and Vivian Ragan, and Gerald Summer's Grocery served our food needs. Saturday night was the day to shop. We even bought pretty dish sets (yellow/white plates with stems of wheat motif) at HyKlas with the coupon rewards from our purchases at the grocery store. Brammer Grocery and Chandler Grocery were smaller grocery stores.

Buzz's Restaurant served many homemade meals and freshly made pies for dessert. They were open on Saturday evenings, and we went every Saturday for dinner. My dad had made a promise to Mother to have one night out as a family. Scalloped chicken was a favorite entree. Cinnamon pecan rolls were my favorite dessert besides a piece of homemade pie. A&W Root Beer Stand was one of the first fast food establishments in town.

Western Auto; Havemann Gambles; and Ted Davis Hardware were the local hardware stores, in addition to Cashman and Lukert Hardware which was one of the oldest hardware stores in town. Lehman and Meyer took care of plumbing needs. George Stollers Interiors and Emert Flooring offered home improvement services.

The Kansas National Guard had an ac-

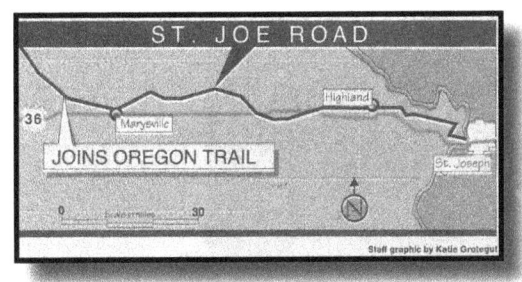

Oregon Trail

Sabetha Water Tower Marking Oregon Trail

tive unit in Sabetha. During the Viet Nam War, the Guard was called up to go to Colorado for training and then on to Viet Nam. My brother, Dan, went to Viet Nam and served as a cook. One of his stories was that the Officers would ask for a case of steaks for two different events. He received two R &R trips for doing this. He went to Hong Kong for one trip and then to Australia for another trip. He was able to see parts of the world while serving in the Guard. The Sabetha Guard had many men leave to serve during this time.

Be Wise Owl Sign

Sabetha Country Club offered golfing, swimming, and social gatherings. Many tournaments and parties were held at the Club. My Mother played golf with a group of gals every week. The Club drew many families to come for meals on a weekly basis.

Sabetha Murdock Hospital, Sabetha, KS

Wenger Manufacturing was founded and owned by Lou and Joe Wenger. The brothers were very inventive and started their business in their garage. Over many years it became a famous worldwide company making extruder machines and many other types of industrial machinery. It is still a large employer in the surrounding area.

There were many other businesses in Sabetha such as Roberts Construction and Garber Cement; Sears; Haxton Barber Shop; Williams Cleaners; McQuillen Cleaners; Dreher Photography Studio; Flott Insurance; Dan Aul, Attorney; Sunny Fabrics; Chiropractors Dr. Cupp and Dr. Kimmel;

Mary Cotton Library, Sabetha, KS

Old Elementary School, Sabetha, KS

and a couple of pool halls such as Eli Hartter Pool Hall and Rokey Pool Hall; and Hatz Cafe where one could purchase a 50 cent meal of a hamburger and French fries.

Mary Cotton Library was also an integral part of the community to check out books. Everyone had a library card.

There were many churches in town: Congregational, Baptist, Apostolic Christian, Church of the Brethren, Methodist, Wesleyan, and Catholic. Rock Creek Church of the Brethren was 5 miles north of Sabetha.

With Sabetha being surrounded by a large rural farming community, going there was the hub of commerce. Towns such as Seneca, Bern, Hiawatha, Fairview and Morrill were nearby. Humboldt, Dawson, and Falls City in Nebraska were in the area also, only a few miles north of Sabetha.

Ernie Block owned the movie theater downtown and the drive-in theater north of town. *Old Yeller, Pollyanna,* and *The Shaggy Dog,* Disney's *Bambi*, John Wayne movies, also *South Pacific*, as well as the other Roger and Hammerstein movies, *My Fair Lady; Sound of Music, Cinderella,* and *Oklahoma*; and black and white movies were a Saturday matinée or night event at the drive-in. As a family, we would go to the drive-in theater on Saturday night. We parked the car, hooked up the speaker that hung on a post, spread out the blankets, and sat through the movie. I really liked buying popcorn and a huge dill pickle.

Chuck and Mary Dornes, Leona, and Bob Murchison are seen in the front row at the theater. Do you recognize anyone?

Eventually, the movie theater was shut down and the building was turned into a bowling alley.

Chuck & Mary Dornes, Sabetha, KS, Theater

My mom and dad joined many bowling leagues, went to state competitions, and won many prizes.

Farmer's State Bank served the banking needs of all of us.

Al Brey Motors, Bill Leman Motors, Gilbert Motors, Sam Cook Ford, and later, Aberle Ford offered the latest US-made vehicles. My first car was a red VW Bug. Brey sold Chevrolets and they didn't know what to do with foreign-made cars; there weren't many in town. Charles Dornes, my dad, was the salesman so he sold the VW to me just to get it off the lot. I paid $600 for the car. I took out a loan at the Farmers State Bank and made payments.

Root Beer Mugs

Sabetha ~~High~~ School (Fountain Villa Nursing Home was built later on the same block)

The Blue Jays at the school had many activities that drew people to town. The old grade school had a metal escape tube that we played on when school was closed. Eventually, the school was torn down and a new grade school was built. The high school gym was busy with basketball games and the football team played on the field behind the high school. After many years, the old high school building was also torn down. So many memories were made there.

High School, Sabetha, KS

Teen Town was primarily for Jr. High Kids. They played spin the bottle and other games while the record player played the latest hits. Kiddo-Lounge came later out on the highway next to Koch's Dairy Queen.

Saturday night was not complete until you bought a freshly popped bag of popcorn from Elmer, the little person; or Clarence Davis in the van, and picked up the latest issue of the Sabetha Herald to catch up on all the local news. Ralph Tennel, Jr. was the editor of the paper.

Chapter 8
James Lane

James Lane

The Jayhawker

Lane was born in Lawrenceburg, Indiana where he practiced law when he was admitted to the bar in 1840. He was a U.S. congressman from Indiana (1853–1855) where he voted for the Kansas-Nebraska Act.

He moved to the Kansas Territory in 1855 and immediately became involved in the abolitionist movement in Kansas. He was often called the leader of the Jayhawkers, a leading Free Soil militant group. After the "Free Soilers" succeeded in getting Kansas admitted to the Union in 1861 as a free state, Lane was elected as one of the new state's first U.S. Senators and re-elected in 1865.

During the American Civil War, in addition to his Senate service, Lane raised a brigade of Jayhawkers known as the "Kansas Brigade."

Lane was the target of the event that became the Lawrence Massacre (or Quantrill's Raid) on August 21, 1863. (Quantrill tried to kill him because of the death of his family in Kansas.) Lane escaped the raid by racing through a cornfield in his nightshirt.

Lane survived many hardships in his life, including fighting in the Mexican American War and the Civil War.

Lane Trail for Underground Railway

THE JAMES LANE FREEDOM TRAIL IS CURRENTLY Highway 75 North of Topeka

Lane's Trail begins in Westport KC leading to Topeka, and winds north through to Nebraska and Iowa. Lane Trail was a safe passageway for immigrants wanting to settle in the area. Plymouth Springs, near the Nebraska border, was laid out by Lane and his army. Mineral Springs was renamed Sycamore Springs in 1886 by two local ladies, possibly because of the many Sycamore trees in the area. Sycamore Springs is within a few miles of Fort Lane. East of Sycamore Springs a small town, Plymouth Springs, and Fort Lane existed.

Cairn or Lane's Chimneys

The towns of Plymouth and Lexington once stood as ports of entry on the Lane Trail near present-day Highway 75 close to Sycamore Springs. The Lane Trail and Fort Lane were established in 1856 to bypass pro-slavery strongholds in Missouri and provide free-state settlers a safe route into Kansas. In 1856, about 600-700 immigrants seeking homes traveled on this trail. Falls City, NE and Holton, KS were founded by these immigrants. Rock piles (much like Cairns) were known as "Lane's chimneys" and marked the trail. Leaving Iowa City, settlers went west into Nebraska and south into Kansas, passing through

Rock Cairns

Plymouth, Lexington, Powhattan, Netawaka, and Holton before arriving in Topeka. The trail also served as part of the underground railroad, used by John Brown and others to transport slaves north to freedom.

At Plymouth (three miles south of the Nebraska line), and Lexington (a few miles south of Sabetha), the settlers built log cabins surrounded by earthen-walled forts for protection. Armed with rifles and bolstered by a small cannon at Plymouth, the settlers established an antislavery presence that helped bring "Bleeding Kansas" into the Union as a free state.

The Lane Trail" Waymarking.com/Kansas Historical Markers (Located North of Sabetha, Kansas on Highway 75)

In the wintry twilight of a January day in 1859, a small caravan of wagons occupied by 30 or 40 escaped slaves approached the log cabin dwelling of Charles Smith in the southwest corner of Brown County. Several out-riders escorted the party, whose leader was the notorious abolitionist, John Brown. This was Brown's last adventure in Bleeding Kansas. A few months later he was captured and executed in an ill-conceived slave insurrection at Harpers Ferry, Virginia—martyred in the cause of emancipation.

Sear fields and bare tree branches bracketed the lonely Smith cabin in stark relief, but hot food and shelter from the elements were a welcome prospect for the exhausted little band on their long journey to freedom in Canada on the Underground Railway.

The Underground Railway was necessarily clandestine (secretive) and an occult (unusual practice) in the Kansas Territory. Slaves were chattels, and those aiding in their escape could be prosecuted for receiving and concealing stolen prop-

Lane Trail Map

erty. Earlier on this particular day, an armed posse barred the slaves from making passage at Dr. Albert Fuller's cabin at the crossing of Straight Creek two miles south of present Netawaka where they had spent the previous night. When no move was made to arrest them, Brown loaded the 30 slaves into wagons and boldly splashed through the ford and up the north bank to find that the posse had fled without firing a shot.

By one account, Brown had stolen the slaves, wagons, and draft animals, and it was their owners who barred the road. The retreat was the better part of valor, however. No one doubted Mad John Brown's resolve to shoot it out if challenged. This incident has since been derisively dubbed "The Battle of the Spurs." They used their spurs, not guns to fight. They traveled north to Powhattan Stage-Coach station area, Capioma, Granada, Lexington; then on to Sabetha, Albany, Plymouth, and Falls City, Nebraska north.

Smiths and Fullers were two of the armed stations established in 1856 by James H. Lane. His purpose was to protect free-state settlers when Missouri River ports of entry into Kansas were blockaded by pro-slavery mobs. There was a cannon positioned near Plymouth. Lane called for a free state army to assemble in Iowa, cross the Missouri River into Nebraska, and enter Kansas near the mouth of Pony Creek in Brown County.

Three hundred armed men under James Redpath, and an additional thirty men led by Preston B. Plumb responded to the challenge. They met and established a settlement which they named Plymouth in the southeast corner of S15 T1S R15E near Sycamore Springs. Plymouth became a post office in 1858 with Morgan Willett as postmaster.

Another station named Lexington was located two or three miles southeast of Sabetha (near the 4-mile corner). E. P. Harris is identified as the proprietor. He abandoned his claim in 1863 at the time of the Quantrill raid on Lawrence. At this point the trail passed due south, paralleling the west boundary of Brown County to Smith's Station, and thence to Fullers. A settlement was made in Calhoun County at present Holton, later to become the county seat of Jackson County. The southern terminus of the trail was Rochester, a hamlet on the Ft. Leavenworth and Ft. Riley Military Road north of Topeka.

Lane's Trail saw little or no immigrant traffic after 1856. The stone cairns on hilltops that served as guideposts were not sufficiently marked to be recorded by the Territorial Surveys of 1857-60. This isolation made it an ideal route for the Underground, and the existence of free-state settlers along the trail guaranteed their safety.

Topeka-Capital Journal Newspaper-2015
Brown v. Board of Education, Sherda Williams,
Brown v. Board site Superintendent

"As a white person, I had considered (slavery) as being long ago and history, but for (a resident whose family members had been free slaves), it's immediate history. I realized the history of slavery and how we treated people echo into lives today."

In the 1850s, white people journeying west stopped at the Sycamore Springs area for fresh water, the natural springs. As stated earlier, the town of Plymouth was laid out by James Lane and his army who gave the mineral waters its name of Plymouth Springs, or Springs, Kansas (also known as Pony Creek). They also established Fort Lane nearby Sycamore Springs in 1856 which was visible until 1883. By 1858 only one house remained.

Lane continued to serve in the Civil War until it ended. When re-elected to the U.S. Senate in 1865, he supported President Andrew Johnson's reconstruction policies, including the veto of the Civil Rights Bill. He lost his support and became despondent and on July 1, 1866, he shot himself and died 10 days later. He was buried in Lawrence's Oak Hill Cemetery.

Sycamore Springs Land, Aerial View

James Lane was valuable in ending slavery with the Underground Railway. The arc of history was bent toward victory.

The Sycamore Springs land became private property when John Downs purchased it from the State of Kansas in 1866, five years after Kansas became a state. It was not until 1886 that the Springs were to be developed for commercial use. John Bougher was the owner of the property at this time. The area became popular for picnics and recreation.

As aforementioned, Mineral Springs was renamed Sycamore Springs in 1886 by two local ladies (legend says), possibly because of the many Sycamore trees and the natural springs in the area.

Sycamore Springs is situated on 60 acres located in Brown County, Kansas in what is known as Pleasant Valley. The land is 7 miles northeast of Sabetha, Kansas and 7 miles northwest of Morrill, Kansas. Legal measurement: Southwest quarter and the west half of the southeast quarter of the southwest quarter of section sixteen, Township one, Range fifteen, (SW1/4 Section 16, Township 1, Range 15).

Pony Creek (John Brown, Abolitionist, hid the ponies therefore the name in the creek when he was moving slaves north to freedom in the 1860s.) The creek runs through the Sycamore land property.

Giant sycamore trees are amid the picturesque hills covered with other forest trees of every description. Natural springs feed from the hillside and run into the creek.

Kansas Flag

Owner Deeds of the Sycamore Springs Land 1851-2020

Mortgage	Date	Grantor	Grantee
Mortgage	Oct 23.1851	Francis W. Starns	Richard L. Oldhaus (Althouse?)
State-Kansas	Sept.4.1866	Governor of Kansas	John W. Downs
Lease	June.10.1869	John W. Downs	School District #10
W. Deed	Oct.6.1873	John W. Downs and Wife	Wm B.Slossom & S. Slossom
W. Deed	Oct.26.1878	WB Slossom & S. Slossom &wives	Edgar L. Clark
W. Deed	Aug.21.1886	Edgar L. Clark	Jacob Brougher
W. Deed	Sept.21.1886	Jacob Brougher & wife	J.W.Scott
W. Deed	Feb.26.1889	J.W.Scott & wife	F.A.Gue
W. Deed	Feb.27.1889	F.A.Gue & wife	John Lanning
W. Deed	Aug.3.1889	John Lanning & wife, Pearl	Trustees United Brethren Church
W. Deed	Aug.7.1890	John Lanning & wife, Pearl	Theophilus Lanning
W. Deed	Feb 25.1893	Theophilus Lanning & wife, Kittie	C.C. Babst
W. Deed	Apr 9.1893	John R. Blanchett & wife	Theophilus Lanning
W. Deed	Sept 27.1892	F.A.Gue & wife	Martha C. Harter (SW 1/4)
W. Deed	Oct 4.1894	Martha C. Harter & husband	William A Griffin
W. Deed	Oct 2.1895	William A Griffin & wife	Enoch V. Kauffman
W. Deed	July 8.1897	William A Griffin & wife	JF Slusher
W. Deed	Sept 23.1915	Henry P. Kauffman	Enoch V. Kauffman
W. Deed	Apr 5.1915	Henry P. Kauffman	JF Slusher
W. Deed	Sept 23.1915	JF Slusher & wife	HP Kauffman
W. Deed	Aug 30.1920	EV Kauffman et al	Sycamore Mineral Springs. SW 1/4
Tax Deed	Sept 12. 1924	County Clerk Brown County	George Ayers SW 1/4
W. Deed	July 7.1927	A.J. Collins Trustee	Clemens Rucker etal SW 14
A.C. Deed	Mar 13.1928	John Koch	Clemens Rucker etal SW 14
W. Deed	Sept 1.1932	Roy R Spring & wife	SM Hibbard-Clemens Rucker 1/3 int.
W. Deed	May 29.1933	Clemens Rucker et al	Sycamore Mineral Springs. Clinic Co.
26 Deed	July 31.1933	J.T. Slusher & wife	Sycamore Mineral Springs. Clinic Co.

Owner Deeds of the Sycamore Springs Land 1851-2020

26 Deed	July 31.1933	J.T. Slusher & wife	Sycamore Mineral Springs. Clinic Co.
Affidavit	Aug 14.1933	T.M. Wherry	Sycamore Mineral Springs. Clinic Co.
Affidavit	Aug 16. 1933	A.B. Lanning	Sycamore Mineral Springs. Clinic Co.
Affidavit	Aug 1.1933	J.T. Slusher & wife	Sycamore Mineral Springs. Clinic Co.
Affidavit	Aug 9.1933	SM Hibbard-Clemens Rucker	Sycamore Mineral Springs. Clinic Co.
Mortgage	May 29.1933	Sycamore Mineral Springs Clinic	Noah Edelman SW 1/4 Released
W. Deed	Sept 15.1937	J.T. Slusher & wife	Noah Edelman SW 1/4 Released
Sheriff Deed	Aug 12.1939	Sheriff Brown County	Noah Edelman
Contract	May 12.1949	Noah Edelman & wife	Charles Dornes, wife, Mary SW-W2-SE4
	June 1.1966	Charles Dornes & wife	Terry & Jerry Tietjens
	Oct 1.1985	Terry Tietjens et al	Thousand Adventures Inc.
	1990	Terry Tietjens etal	Dale & Betty Aue
	2020	Dale & Betty Aue	Kent & Molly Grimm

Chapter 9
Sycamore Springs

Mr. & Mrs. Jacob Bougher

Sycamore Springs
Private Owners
1886-1920
Jacob Bougher, Owner
Sycamore Springs-1880

Sycamore Springs was first owned and operated as a health resort and public picnic area by Jacob Bougher in the early 1880s.

In 1886 the merits of the water became known to early settlers. Indians and pioneers with prairie schooners stopped there for days at a time to rest and drink the water. The Indian medicine man had already discovered the value of the water. At this time the trail going from the West to St. Joseph, Missouri passed through the springs' farmland. Heavy timber and thick underbrush surrounded the stream (Pony Creek).

Mr. Bougher was the first man to give baths to the public.

His bathhouse and tub consisted of a hole dug in the ground, 5 feet deep, near the springs, filled with water.

The following articles are copied from the archives of the Brown and Nemaha County area newspapers.

July 30, 1886
Brown County Herald (Morrill)
L.W. Phillippi
VISIT SYCAMORE SPRINGS

Among the famous early explorers of North America was a Spaniard named Ponce de Leon, an aged man in disgrace, who believed that there somewhere existed a fountain wherein, a person might bathe and become young again. This fountain, the age Spaniard believed, was situated somewhere in the Southern part of North America, and in his search for it he became the discoverer of Florida but in Brown County, Kansas, five miles northwest of Morrill and Jacob Bougher's farm.

It has long been known to the people of the neighborhood that the waters of this spring contained certain medicinal properties, but it had never gained publicity until this summer. That the water is of a mineral nature there is no doubt, as those who testify to its exhilarating effects. The worst afflicted invalids, who found relief nowhere else, declare that a week's camping at Sycamore Springs and drinking the water and bathing is worth more than all the medicine they have taken.

The spring is beautifully situated in a fine grove on Pony Creek and a more pleasant resort can nowhere be found. The vicinity as yet has somewhat the appearance of a wilderness, but workmen are putting up bath houses, dancing platforms, etc. and it will soon have the appearance of a health and pleasure resort.

Daily can be seen scores of people there from Sabetha, Morrill, and other places, some of whom stay there all the time, finding a cool and pleasant place for health and recreation.

There is no reason why this spring should not become as famous as any other in the country, and all it needs is proper effort in booming it, for the beneficial effects of the water area sufficient testimony of its remedial virtue.

I would advise the disordered and afflicted to visit this fountain, drink of and bathe in its limped waters and restored to youth and health.

On Saturday, July 31st., there will be a bowery dance and picnic at the Springs. Everybody is invited to attend and a large turnout is anticipated. Arrangements are being made which will insure unlimited enjoyment to everyone present.

August 6, 1886
Brown County Herald (Morrill)

"The following rules are posted up at Sycamore Springs and Mr. Bougher says he will enforce them: No drinking and carousing. No hollering or loud talking Respect must be shown to ladies. No breaking and destroying the timer. Any person hitching to trees will find their horses loose. Positively no person must hitch to trees."

August 6, 1886
Brown County Herald (Morrill) Old Nick
PONY CREEK

"The Mineral Springs are all the rage on Pony Creek. The last few days teams can be seen coming from all directions bound for the Springs. We visited the Springs on Saturday evening and Sunday for the first. On Saturday the ground was well filled with both old and young, and all seemed to enjoy themselves. The young danced, while the old looked on. Sunday morning people began to gather in and kept coming until everybody was there.

"Mr. Bougher is putting forth every effort to make the Springs a place of pleasure, for both old and young. He has his bathrooms now in good shape and is ready to accommodate you at all hours of the day. On Sunday those rooms were occupied constantly. The following named people can testify as to the effects of the water: Al Lyman, James Watts,

Dennis Wymer, J.W. Watts, Jake Hoffman, Charles Sprague, T.P. Gordon, George Adams, J.F. Clugh, Mr. Holtzshoe, A.C. Moorhead and many others."

August 7, 1886 Sabetha Herald

"The rush to Sycamore Springs continues. At last, Mineral Springs has received a name and will henceforth be known as The Sycamore Springs.

There will be a meeting held at Sycamore Springs on Sunday, August 8, at 3:30 P.M., under the auspices of the YMCA. The Sabetha delegation will start from the Methodist Church at 2 p.m. Where conveyances will be waiting for those who wish to attend."

August 13, 1886
Brown County Herald (Morrill)

"Sycamore Springs is becoming a great resort for those who have been so unfortunate as to be attacked by disease, and pleasure seekers. Sunday there were 40 double and 36 single teams (horse teams) admitted to the grounds besides several hundred footmen. The bathtubs were in constant use. All who use the water pronounce it superior to any they have ever used."

June 14, 1888
Sabetha Herald
SYCAMORE SPRINGS (HISTORY)

Before the 1860's when this country was only known by the few hardy pioneers who had log cabins along the streams, it was known that the few women who lived on Pony Creek, Northeast of Sabetha, by common consent, would meet at least once a week around a clear running spring, and there, without soap or toilsome rubbing, do the week's washing in a few moments by merely dipping or wringing their clothes a few times in the waters of this spring.

Later, about 1865, one of Albany's early settlers, sorely afflicted with rheumatism camped a few days on the banks of this stream and bathed in and drank the water of this

The Gatekeepers of Sycamore Springs | 75

spring, resulting in a speedy cure.

Later, under the name of "Sycamore Springs" given it by one of Sabetha's fair ladies, Miss Ella H. Taylor, these springs have become quite a famous watering place for Sabetha, Northern Kansas and Southern Nebraska.

Later, under the name of "Sycamore Springs," given it by one of Sabetha's fair ladies, Miss Ella H. Taylor, these springs have become quite a famous watering place for Sabetha, Northern Kansas and Southern Nebraska.

An analysis of its waters made at the laboratory of the Brown Medicine and Manufacturing Company, of Leavenworth, shows it to contain magnesium, iron, sodium, chloride, potassium and sulfur. Many cures are reported of rheumatism, scrofula, dyspepsia, also of blood and skin diseases.

Those who have bathed in the noted springs of Poncha and Manitou of Colorado, declare the baths of Sycamore Springs to be far superior in every respect. As you recline in the bath the pleasant sensation of being surrounded by a soft, almost oily covering of pure clear water, gives an exhilarating sensation, to the worn and tired patient, or to one who requires only rest and recreation from toil or business.

During the last three years, since accommodations for invalids and pleasure seekers have been provided, during the hot summer months, the place is crowded with visitors and hundreds are found camping beside its life-giving waters.

Among the many attractions which Sabetha offers for home or pleasure seekers, Sycamore Springs will, in a few years, be recognized as a pleasure resort and sanatorium equal to any in the West.

August 16, 1889 Sabetha Herald
The Flood At Sycamore Springs

The storm that swept through our vicinity last Monday night, took in the Springs as well. And as that gave evidence of being the most exciting event of the afternoon, the party of tenters and boarders gathered on the piazza of the hotel to watch the rising of the water, in the creek below.

In less than an hour the water had risen over ten feet and covered the bridge soon after the party of tenters had crossed it, to seek shelter under the hospitable roof of the hotel.

By 5 o'clock the water had spread all over the flats and was carrying down heavy logs and brushwood at a rapid rate.

All were watching the little bridge, which was the last connecting link between us and the tents and provisions on the other side. It held out until just before dark when it gave way and floated down the stream.

About 12 o'clock the heavy shower began to make a deeper impression upon the creek, for within five minutes it had risen above a piece of ground that had been high and dry before; and the stream then was nearly a quarter of a mile in width.

During the vivid flashes of lightning pieces of broken bridges could be seen floating rapidly by, and the prospect of home-returning looked rather dubious. But with the morning came the sun, and the water had begun to recede rapidly.

So with the sunshine, a palatable breakfast and the arrival of a friend who reported the roads passable, the prospects brightened, and all decided that a quiet ordinary day at Sycamore Springs would not have been half so interesting as the one just experienced.

The Sycamore Springs guests came home bag and baggage Tuesday to await the drying-up process.

F.A. Gue, Owner Sycamore Springs 1889

F.A. Gue was a Union soldier during the Civil War. As Assistant Surgeon, he was captured while taking care of 38 Union soldiers. He spent 526 days in prison during the War of the Rebellion. He was at Libby, Pemberton, Danville, Andersonville, and then Salisbury, N. Carolina prison. The Salisbury prison, merely a stockade, burned down and Gue was taken to Florence, South Carolina where he was kept until near the close of the war.

The health of Gue was in a miserable state by this time. He settled in Brown County, Kansas near Sun Springs where mineral water was located. He purchased Sun Springs in the 1870s and started to build a sanitarium but soon sold it.

Several years later in 1889, Gue purchased Sycamore Springs. He had come to the area for health treatment and saw the value in the mineral springs and how the water would help any inflammation.

One of the Springs 1896

In 1889, Gue along with Jake Hartter built the first hotel—3 stories, 40 rooms accommodating 30 people. Gue also built the first bath houses, 8x10 feet with a tub. Water was heated outside in a large kettle and carried into the bathhouse. A patient would plunge into the cold water and then into the warm water. The mineral water became well known in eastern Kansas and surrounding states in the 1890s, and many people stayed at the hotel to rest and drink the mineral water. Victims of chronic ailments journeyed to these marvelous gifts of nature and found in the spring water the power to restore them to health. Neighbors took in roomers so that everyone might have a place to stay.

Hartter built a dam in the creek to hold the water. He also built a big ice-house that had a storage capacity of 100 tons of ice. C.C. Babst and Frank Miller packed the ice each winter.

July 11, 1889
Sabetha Herald
THE FOURTH AT SYCAMORE

Having a few hours of leisure on the Fourth in which to celebrate. The Herald editor took a piece of bread and cheese, a hammock, and a copy of Robert Eismere and started for Sycamore Springs, reasoning that "far from the madding crowd" nothing would interfere with an afternoon nap in the silent shade of this sequestered spot.

Judge then of the surprise experienced upon reaching this quiet retreat, to find about two thousand souls "reckoning," as Mark Twain would say, a soul to each person, on the grounds already, evidently with a different view of the case. The small boy was there, likewise the large firecracker, refreshments were on hand and the land appeared to be flowing with ice cream and lemonade.

Three Story Hotel

The hammock was in requisition; but "Robert" was left, and although displayed in all the brilliancy of a showy cloth lining on the top of the buggy seat nobody could be induced to steal him.

Mr. Gue, the present proprietor of the Springs, has made solid improvements and it is easy to see that an immense amount of work has been done. The hotel, a three-story structure, is very picturesque with its verandas and clean, fresh, airy rooms and would make a delightful place for rest and recuperation. We would recommend it as a relief to the jaded nerves and heated head of the Kansas City Times, only there is no bootlegging allowed about the premises, and the strictest order prevailed, even with the immense crowd gathered there last Thursday.

Water Tower on the Hill,

In the hotel, the spotless linen of the dining room and the absolute cleanliness of everything showed that a notable housekeeper holds the reins of the inside management and that the comfort of the guest is of paramount importance. Mr. Gue kindly showed us the rooms, and a number were immediately engaged by Sabetha parties for occupancy in August. The hotel is a vast improvement on tent life after flies, mosquitoes, and other proboscises insects begin to take an active interest in affairs, and if Mr. Gue had twice as many rooms, they would all be occupied.

Spring 1896

February 24, 1916
Sabetha Herald Sycamore Springs Burns

Sycamore Springs, the most noted health resort in Northeastern Kansas burned Tuesday afternoon. The big five-story stone building containing sixty rooms is in ruins, and the old frame building just south of it is in ash. The loss is estimated at $25,000. The Herald is unable to learn the exact amount of insurance, but it is believed to be somewhere around $10,000. E.V. Kauffman says he does not know as yet whether the building will be rebuilt, but that the sale of the mineral water will continue.

The fire was discovered about 3 p.m. Tuesday. It was high up on the north end of the old frame building. The modern building was just north of the old wood building.

Mr. Kauffman believes that the fire started from the chimney and that it had been smoldering all day. The Springs have a reservoir on the hill but

E.V. Kauffman, Proprietor

Sycamore Mineral Springs Brochure Ad

it didn't have much water in it. Mr. Kauffman was the only one at home. Neighbors were quickly summoned. Joe Slusher came from a sale nearby. Soon fifty men were fighting the fire to keep it from spreading from the old building to the new. It looked as if the crowd would win. Water was dashed on the new building, and by willing hands, men worked heroically.

Finally, it got so hot that the men could not work any longer. The fire caught on the cornice of the modern building and in a few minutes the flames were beyond control. Much of the furniture was saved, but it was badly damaged in being gotten out of the building.

Mr. Kauffman and Joe Slusher have moved into the Springs cottage in the woods and have established headquarters at the pop house.

The Springs have been used for mineral water for 30 years. Mr. Kauffman has been there for 20 years.

The Atchison Champion Newspaper, May 19, 1900, Article excerpts
The Sycamore Mineral Springs—A Health Resort is fast gaining in public favor

In the last few months, the proprietor,

Sycamore Mineral Springs Postcard of the Hotel

E.V. Kauffman, has added greatly to his facilities for making guests and patients comfortable. Last fall a new steam heating apparatus was installed, and new bathrooms with the latest and most convenient fittings were built. The part of the buildings formerly used for bathrooms has been rearranged and converted into additional bedrooms and the dining rooms have also been enlarged. The whole building has been thoroughly overhauled and renovated and numerous conveniences added. Mr.

Hotel Early 1900s Newspaper Photo

Kauffman will open the season with facilities for taking care of nearly twice as many people as heretofore.

One feature of the business is the bottling and shipping of the spring water. Many people derive great benefits from drinking spring water. The cost of giving the water a trial in this way is but slight and few who try it fail to be benefited. It is put up in 5-gallon carboys or demijohns (a bulbous narrow-necked bottle holding from 3-10 gallons of liquid, typically enclosed in a wicker cover) and also in cases of quart bottles. This branch of the business keeps one man busy most of the time and the demand is rapidly increasing.

Enoch V. Kauffman, the postmaster at Springs, Brown County, Kansas, was born Nov 6, 1840, near the banks of the Shenandoah River, Page County, Virginia. He emigrated west in 1867. He married Mrs.

Sycamore Mineral Springs Side Entrance

82 | Bonnie Dornes Hanni

Hotel Burned 1916, Sabetha Newspaper

Sarah L. Slusher, daughter of Joseph Harter, of Hancock County, Illinois, on Feb 2, 1869. They came to Oneida, Kansas, in 1881 and to Brown County, Kansas in 1895. They settled at Sycamore Mineral Springs, the great health resort of the west, where his wife and daughter, Miss Susie Kauffman, still reside. He has 3 children living: Henry P. Kauffman, Lincoln, Nebraska; William L. Kauffman, Bern, Kansas; and Susie Kauffman.

The people of Brown County take great pride and satisfaction in the growing use of the pleasant resort and being perfectly familiar with the merits of the water as a health-giving agent. And expect to see it grow as the years pass by until it is second to none in the west.

Women by the Spring

Mr. Kauffman is constantly receiving unsolicited testimonials from people whose sufferings have been recovered by the use of the water. He deserves great credit for his energy in enlarging the capacity of the house, and no doubt his enterprise will be rewarded by the largely increased patronage.

It is estimated that 125,000 gallons of bottled water are sold to

Four Story Hotel

the public each year.

As mentioned in the brochure "Water Way to Health," Sycamore Springs Sanitarium was not a hospital or sanitarium as these terms are most used and understood. It was, in fact, a very pretty, homelike resort hotel, operated as a hotel, with separate departments designed and equipped for the baths, Swedish massage, x-ray, and Colonic treatment of value in correcting chronic disorders.

Hotel with Bridge Postcard

Patrons had the privilege of taking as much or as little of the service as they needed. Some people needed only room, board, mineral waters, and baths, while others require the entire institutional course of treatment in addition, which could be had for a very little extra cost.

Sycamore Springs was sold in 1895 to Enoch Kauffman (United States Postmaster) and Joe Slusher probably its most important promoter. Besides becoming well known for its mineral springs, the recreational park of Sycamore grew rapidly. A large swimming pool was built in 1908, as numerous cabins, a ballpark and dozens of Penny-land carnival concessions stand, and other types of carnival entertainment. All ages enjoyed the merry-go-round powered by a stationary steam engine. Many enjoyed the dance floor or a concert from the bandstand.

Admission was charged at the gate. It became a gathering place for many families.

Pennyland was under the management of Mr. Elrod of Lincoln, Nebras-

Hotel Burned 1916, Sabetha Newspaper

ka. He suddenly disappeared with a very large sum of money and left many unpaid bills. For many years after that, the swimming pool was the only recreation there.

Guests on the Hotel Porch

Old Spring Postcard

Pennyland

Chapter 10
20th Century

Carousel Merry-Go-Round

1902-1935
Kansas-Nebraska Railroad
Sycamore Springs

On September 25, 1902, the Sycamore Railroad and Improvement was duly organized under the laws of Nebraska having a capital check of one million dollars to be known as the Falls City, Sycamore Springs, Sabetha, and Southwestern Railway Co. It was to build a standard gauge railway from Falls City to Sabetha, Kansas by way of Sycamore Springs and their intention was to take over the springs and build a large hotel and lake and other extensive improvements, making it the ideal health resort.

Kansas Corporation Seal for Railroad

Post Office Replica

They received the Kansas Corporate Seal, but the railroad was never built and only on paper. The coming of the auto and the high cost of land rights-of-way put an end to this railroad dream.

E. V. Kauffman was the prime mover for the railroad. He had the Kansas spirit, an entrepreneurial spirit to promote the Springs.

A Post Office was established in 1858 at Plymouth with Morgan Willett as the first Postmaster.

From 1886 to 1902 Sycamore had a United States Post Office known as Springs, Kansas and was discontinued in 1902.

Natural Spring Water for Sale

An important part of early Sycamore history is the bottling and selling of crystal-clear spring water. Many can remember the selling of five-gallon bottles of water at fifty cents. This continued into the 1940s-the 1950s.

In the 1940s Langdon Livengood worked for Sycamore Springs. He worked with Mr.

Chapman and bottled water that was to be sold.

In February 1901 Kauffman began producing bottled pop. Various flavors were mixed with the mineral water making it a very satisfying drink.

Trucks were used to deliver mineral water to people in neighboring towns.

A general store existed at the time, and many improvements were made at the hotel. Many people benefited from the attributes of the natural spring water.

In 1904 a modern rock ho-

Water Delivery Truck

tel was built containing fifty bedrooms. It was truly ultra-modern in its day, heated by steam, and hot and cold running water with modern hospital facilities. As noted earlier, a fire on February 22, 1916, destroyed both hotels.

In 1917, parts of the structures were rebuilt and are now the present-day hotel.

Sabetha Doctors Purchase
Sycamore Springs
1920

On August 20, 1920, 2 doctors, Dr. S.M. Hibbard and Dr. Clemmons Rucker; and 2 investors, Roy Spring and John Koch from the Sabetha, Kansas area purchased the hotel and springs.

World War I (1914-1918) came to an end in 1918. Soldiers returning from the war came to the Springs for rest and recuperation.

The hotel had partially been rebuilt in 1917 to what we see today. The sanitarium was to provide for the care and treatment of the health seekers as well as all manner of pleasure and entertainment. No treatment was given for contagious or infectious diseases, all were non-surgical treatments.

Gathered Around the Upper Spring No. 2

J.T. Slusher operated the recreational part of the springs. Admission to the park was 10 cents. Many concessions and kinds of recreation have been popular over the years. All ages enjoyed the merry-go-round powered by a stationary steam engine. Many enjoyed the dance floor pavil-

View of the Hotel & Doctor Clinic

The Doctor's Clinic

ion or a band concert from the bandstand. Pennyland concessions, free vaudeville, tennis, fortune telling, a shooting gallery, and go kart racing track, and riding on the auto coaster added to the fun for everyone there. One of the most popular areas of the 1920s was the baseball diamond, not to mention the swimming pool which was built about 1920. No dancing was allowed on Sundays.

Doughnuts, popcorn, peanuts, hamburgers, and ice cream were available with coffee, Root Beer, and soft drinks.

Water Therapy

Water therapy is also called hydrotherapy and may be restricted to the use of aquatic therapy, a form of physical therapy, and as a cleansing agent. Hydrotherapy involves submerging all or part of the body in water. This type of therapy first appeared in the United States in the 1840s from Europe. Water-cure establishments were mostly located in the northeast part of the United States. Since Kauffman came from the East, this was probably influential in his decision to purchase after learning of the water at Sycamore Springs.

The hotel had an area on the first floor where the showers and tubs were located.

As a kid, I remember going into the hotel and seeing the dressing room cubicles and the place where the showers were located. It was very cold and damp in the main room. The room was on the side of the hill and the limestone walls kept the room cold. The room was turned into a living room

Partially Rebuilt Hotel

later when the Tietjens operated the hotel "I was always afraid of going upstairs because I felt there was a ghost presence at the bottom of the stairs. Who knows, maybe someone forgot to leave the sanitarium."

In the Dirty '30s, 1929-1940, Kansas was severely affected. Farmers suffered the most. Dust filtered into every building, home, every pore of the human body, and most hazardous, into the lungs of thousands of people in 1935, much like in 2020.

Samuel M. Hibbard, Doctor

President Franklin D. Roosevelt implemented federal programs, but hundreds of acres of topsoil were already lost and unfit for cultivation. The drought didn't end until the rains came in 1939.

Dust masks, and even gas masks, became everyday wear (wearing masks in the 2019-2021 Covid Pandemic had much the same feeling). Wet sheets were hung in windows to keep out the dust, and farmers watched helplessly as crops blew away. Stock water tanks were filled with dust, and implements were buried in the dust while it piled up in ditches. Animals died from dust inhalation, and lack of water and feed added to the devastation.

Clemens Rucker, Doctor

April 14, 1935 was a day the sun didn't shine in most of the western half of Kansas, known as Black Sunday. It was also during the Dirty 30s, the Sycamore Springs Sanitarium became an even more important place to attend to illnesses caused by dust and the mental well-being of the patients.

Economic devastation hit the nation also. The WPA (Workman's Progress Administration) and Civilian Conservation Corps put men back to work. Many of the bridges in the area were built by the men employed by the WPA.

Conservation of the land became a high priority. The Shelter-belt Project called for the planting of native trees forming the windbreaks, conserving the topsoil from blowing away.

Hiawatha World Community Spotlight excerpts June 24, 2014, article "The Dust Bowl-an ill wind that blew no good."

The Dirty 30s, the years between 1929 and 1940 created a wide swath up the middle of the nation, choking waves as the Dust Bowl swept from Texas to Minnesota. In 1929, the stock market crashed causing the Great Depression until WWII.

Roy Spring, Investor

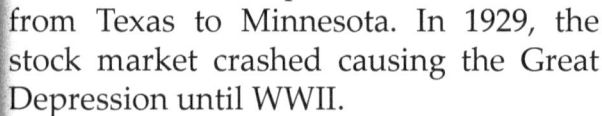

The dust began to blow in huge clouds in all directions for 8 years. The drought that helped bring on the Dust Bowl caused economic disaster in 27 states. Midwest families loaded everything they could on their vehicles and headed to California. Handbills posted in the hard-hit areas promised good jobs but were empty promises and California was just as devastated as the Midwest.

John Koch, Investor

President Franklin Delano Roosevelt began building Federal programs to help the people. But the loss of topsoil started in 1931 and ended in 1939 when the rains finally came.

Works Progress Administration (WPA) and Civilian Conservation Corps (CCC) were formed to put men back to work and to restore or save croplands. The Shelter-belt

Owners by Spring: Roy Spring, Clemens Rucker, John Koch, Samuel Hibbard

**Sanitarium Newsletter
Left Side**

**Sanitarium Newsletter
Right Side**

Project called for the planting of native trees in the Midwest section of the country forming windbreaks.

Today one can still see the remnants of the windbreaks in the countryside. Farmers are removing the Osage Orange hedge rows (Hedge Balls trees) today to increase the amount of land they can cultivate. Can another drought occur? Can we experience another Dust Bowl? Conservation techniques are key to keeping the topsoil. April 14, 1935, was a day the sun didn't shine in most of the western half of Kansas, known as Black Sunday.

During the Dirty 30s, the Sycamore Springs Sanitarium became an important place to attend to illnesses caused by dust and the mental well-being of the patients.

**Putting Final Stucco on
Buildings, 1930**

Sycamore MIneral Springs Postcard

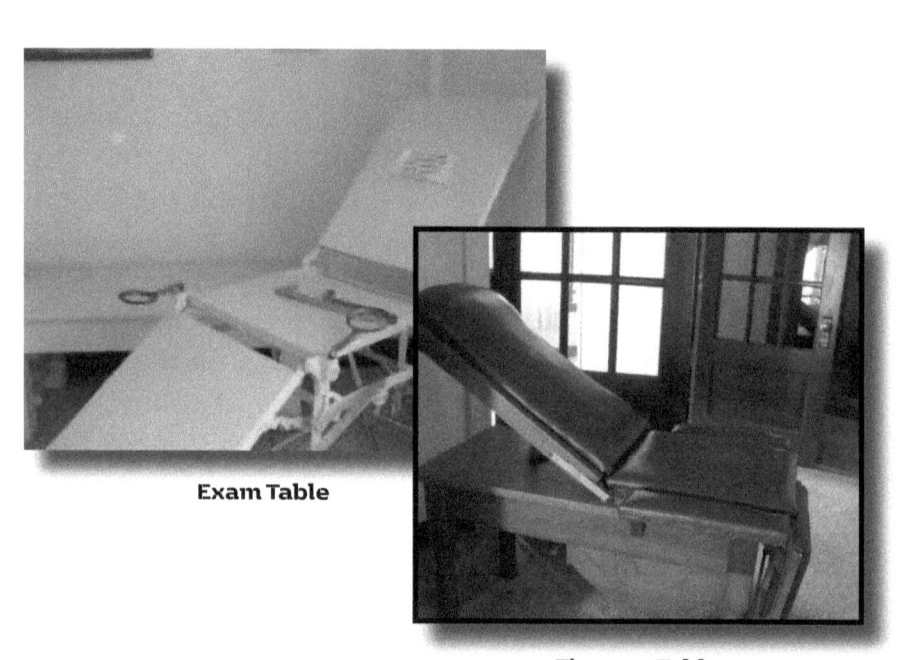

Exam Table

Therapy Table

Hibbard Brochure Water Way to Health

Example ofo Water Therapy Bath House

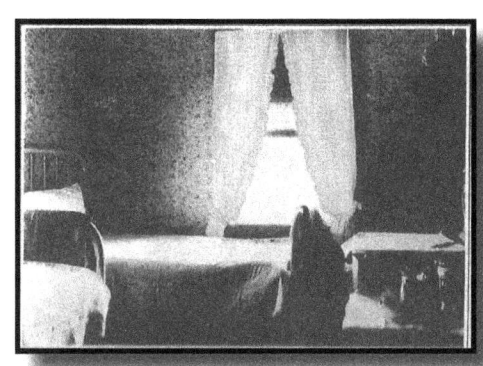

Hotel Guest Room Health

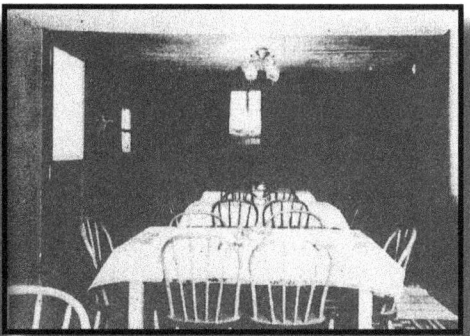

Hotel Dining Room

Cabins for PatientsHealth

**Sycamore Springs
Sanitarium Postcard**

Black Sunday in the Midwest

Kansas Farm Dust bowl

Dust Bowl Masks

Laurel Slayton Family—Bonnie Dornes Hanni Grandparents during the Great Depression

Chapter 11
1935 - 2017

Noah Edelman

Noah and Lillian Edelman,
Owners Sycamore Springs 1937-1949

Noah Edelman and his wife Lillian purchased Sycamore Springs in February 1937 from the Sycamore Mineral Springs Company. They had two children: Nila, 3 years old; and Morris born in one of the rooms at the hotel in 1937. Morris is the only child born at Sycamore Springs.

Noah Edelman, The New Owner of Sycamore Springs, was an entrepreneur, ambitious in improving facilities. He found the bathhouse and pool in poor condition. With his "spunky German work ethic" (Nila Edelman Snyder, daughter quote), he rebuilt it.

At this time, the spa part of the springs took second place, and the water was used to fill the pool only. But in 1939, a skating rink was added. The skating rink was built in 1939.

WWII (1939-1945) came along and doctors and nurses who were working in the sanitarium were called into military service. Edelman closed the sanitarium in 1941 and turned the place into a resort for campers, church groups (Camp Kanebwa for Kansas, Nebraska, and Iowa churches), and boy and girl club groups.

The hotel and shelter house were used to house the camp groups.

At that time, the shelter house was located behind the skating rink. This building was the former bath house attached to the swimming pool area. It was moved farther back behind the skating rink and new concrete block dressing rooms were built.

Lillian Edelman & Nila & Morris

Camp House

For a few years, a café was started in the main house, and located in the re-purposed basement with red booths and square tables and chairs. There was also a full kitchen in the basement that was used for the café.

Howard Kreitzer from Sabetha came home from the war and became a cook during this time, and it was operated as a steak house. Nila was only 15 years old but served as a waitress.

Caroline Edelman from Sabetha was a cook along with Kreitzer in the café.

Mary Slayton Dornes lived at Sycamore as a "hired girl" and worked in the café in 1944.

Camp House with Girls

Cafe Entrance

During the times the camps were at Sycamore Springs, Mrs. Noah Edelman, Lillian, cooked three meals a day for the groups that stayed in the hotel and camp house. She was not involved in the steak house. She raised all the vegetables, and the meat came from their livestock. In the summer, she had one or two hired girls to help clean and cook. One of the campers' favorite meals was her fried chicken, potatoes, homegrown corn on the cob, tomatoes, homemade rolls, and always some kind of dessert.

Cafe House Entrance

Noah Edelman raised cattle at Sycamore. Standing in Pony Creek

Nila Edelman Snyder recently spoke of her parents, Noah and Lillian Edelman, and their life after selling Sycamore Springs to Charles and Mary Dornes in 1949.

Nila said, "Dad and Mom, after leaving Sycamore, built a new home on a farm on Pennsylvania Avenue, about 4 miles east of Bern, Kansas. They lived there until they retired and moved to Sabetha in the 60s. Mom died in 1974, and Dad passed away in 1983.

"Morris, my brother, farmed the family farm, married Marge Menold, and is now semi-retired, lives close to Sabetha.

"As for me, I married Leland Snyder, whom I met at Sycamore where so many other couples found their mate. We lived on a farm north of Sabetha for 63

Cafe House with Nila & Friends

years. Leland passed away in 2016. In 1973 I purchased the Gift and Jewelry Shop (we called it "Nila's") in Sabetha, enjoyed it so much, and closed it in 2011 and retired in Sabetha."

Recreation at Sycamore Springs

Edelman hired young high school boys to lifeguard at the pool and help keep the grounds clean. The grounds were full of people having picnics and family reunions every weekend. During the war, folks from Kansas and Nebraska did not have the gas or tires to travel far, so they utilized Sycamore Springs as their entertainment. Many of them met their future husband or wife while skating or swimming.

Horses at Sycamore Springs

Noah's Cattle in the Creek

The second half of the pool was added by Noah Edelman. This picture shows the swimming pool entrance before fresh crystal-clear spring was added.

The pool was not surrounded by a fence at this time. The sand was placed around the pool before cement and a high fence were installed. Gene Stapleton was one of the lifeguards. Edelman added the second half of the swimming pool, and a wide section of cement surrounded the pool, and a fence was added to enclose it.

Empty Swimming Pool

August 6, 1947
The Sabetha Herald
"Carp fish in the swimming pool"

Carp is still at large. Despite a large crowd of swimmers at Sycamore Springs on Sunday, most of them intent on capturing it, a 10-pound carp continues to roam about the depths of the swimming pool there. A prize of $5.00 was offered to the person who caught the fish by hand. One swimmer was reported to have had his hands on the big fish but it slipped away. A group of boys cornered the carp in shallow water at one time and it looked like success was in sight, but the wary old denizen of the deep eluded them and swam hurriedly away.

Nila Edelman Snyder by the Pool

"The offer still holds," said Noah Edelman on Monday, "and we may have another contest with the same fish next Sunday. In fact, it looks like we may have to give the swimmers a handicap by narrowing the field of activity." The big fish not only doesn't bother swimmers but seems to have a knack for being a hermit. Lawrence Fletchall, who operates the pool concession, is keeping a sharp lookout for persons with fishing tackle and dough balls.

Swimmers in the Pool

The Spring

Edelman built ½ of the skating rink in 1939 from rough-hewn wood and added the second half of the skating rink in 1940. It measured 160 ft. long by 55 ft. wide, the largest skating rink in Kansas.

Every person used clamp-on skates. Many parties were held at the rink.

Every Halloween there was a big party. Games were played while on the skates. There was always a tug-of-war with a huge, long rope between the Nebraska kids against the Kansas kids. Lots of laughing and fun happened that night.

Charles Dornes, who liked being called Chuck, was a "skate boy" in 1941. He would help skaters put on and adjust the clamp-on skates.

The Spring

Chuck Dornes
"Skate Boy"

The Skating Rink

Clamp on Skates

Here are a few memories from Nila Edelman Snyder about her dad, Noah Edelman:

"There was a group or gang of older boys from a neighboring town, the river rats, who came skating quite often and always caused trouble. Dad loved most of the kids that came to skate or swim but these guys were a problem, tripping others on skates, skating too fast and pushing others, causing other skaters to fall, and worst was they stole some skates.

"Dad was not a large man, little but mighty in many ways. Dad caught them stealing a pair of skates. The guy was twice the size of dad. Dad literally threw him out of the rink door, taking part of the door sill with him.

Camping Brochure

"Another time, the same gang of boys from a neighboring town, decided to hang around after the skating rink closed. 'Noah heard one say let's stick around and have some fun.' Dad always carried the money bag with the day's receipts and walked home across the bridge to the hotel and into our living quarters.

View of the House from the Bridge

Walking from the skating rink was always so dark, no lights were lighting the way, and the trees shut out any moonlight.

"Now, this evening, Dad had heard the guys talking, and he decided to take his 22 shotgun

View of the Bridge

(which he had hidden under the counter at the rink) with him while walking home. One of the guys came up to him and asked Dad if he was going to shoot ducks. Dad replied, 'You never know what I may do.' This bunch of guys never bothered him again, at least we never heard about it after that incident."

Nila still has more stories that she remembers. She recalls so many wonderful memories she had while growing up at Sycamore Springs.

Charles and Mary Dornes, Owners Sycamore Springs 1949-1966

In 1949 Charles and Mary (Slayton) Dornes purchased Sycamore Springs for $35,000.00. They sold the Guernsey cows they were raising as a down payment to make the purchase. Within two years, they received the final deed to the land.

"A new bridge was installed as shown above. My grandmother Martha Slayton read in the Falls City Journal newspaper that there was an army bridge for sale. We needed to replace the wooden bridge. She called it "her bridge" whenever she came to visit us. The bridge was originally to be used upside down, but my dad turned it over and used the sides for safety car crossing."

They had two children: Daniel (Danny), 3 years old; and Bonita (Bonnie) 3 months old when they moved from the farm to Sycamore. They operated it as a recreation center for 17 years.

My mom and dad were very good skaters. They would almost float around the rink. Dad wore a whistle to keep control of the rink floor. He would make the guys who were chasing each other slow down and not cause others to fall.

Our Home

Danny & Bonnie Dornes 1949

Danny & Bonnie 1950s

Danny & Bonnie 1955

Charles & Mary Dornes at the Skating Rink

Charles & Mary Dornes, new owners of Sycamore Springs. Charles is 31 years old, and Mary is 23 years old.

106 | Bonnie Dornes Hanni

Originally a shop for Chuck Dornes. He sold Massey Ferguson and Massey Harris machinery in Hiawatha with the Shockleys. He repaired farmer's equipment in the shop.

The Shop 1959

Dornes Hall

Chapter 12
Our World in turmoil

John F. Kennedy Cover,
Life Magazine 1963

Korean War—1950 - 1953
Vietnam War-1955 - 1975

Life and Times of Charles Dornes and Mary Dornes

My great-great grandmother Abbie Dunn married George Slayton, parents of Laurel Slayton. Abbie's father, my great-great grandfather Ransom Dunn, was the founder and President of Hillsdale College in 1844.

Hillsdale College is one of the only schools that does not accept federal funds to maintain the college. It is a conservative liberal arts college. The family learned of Nebraska land from the Nebraska City Arbor Lodge Morton family. The Dunns purchased land in Nebraska as a way to retreat for rest. My grandparents lived and raised their family on the Dunn farm near Salem, Nebraska.

Charles Henry Dornes was born on January 12, 1918, the son of Henry Bert and Emily Molly Martensen Dornes in Alta Vista, Kansas. In 1923, when Charles was 6 years old, he moved with his family to a farm near Oneida. He often recounted his special memory of getting to ride with his dad as they moved the family's possessions in the wagon pulled by their team of horses, Dick and Dan. Growing up, he attended Summit School (now located in Albany) before the family purchased Pineview Farm in the Sycamore community. He remained on the farm to help his parents and also worked with Keith Van Horn, tending his herd of Guernsey dairy cows.

Chuck Dornes 1955

As a young man, Chuck enjoyed spending time at Sycamore Springs and worked there as a "skate boy." He was a quiet and shy young man which presented a challenge to a pretty young thing from Salem, Nebraska, Mary Slayton. My mother, Mary Slayton came to Sycamore Springs for recreation and worked for Noah Edelman as a hired girl, helping with the camps. It was at Sycamore Springs that she met Charles Dornes and the rest is history.

Chuck finally got up the courage and asked her on a date. We are told it took him 6 months before he was brave enough to ask for a kiss.

He was united in marriage to Mary Louise Slayton on October 29, 1944 at the Slayton family home in Salem, Nebraska. Mary and Chuck were the first couple married by Rev. Dennis Kesler, pastor of the Rock Creek Church of the Brethren, and they remained active in the Rock Creek Church.

They lived on the Henry Dornes family farm east of Sycamore Springs (close to Plymouth Springs which James Lane founded) where my father took up dairy farming, following receipt of a registered Guernsey bull as a wedding present from Helen and Keith Van Horn.

Two children, Danny and Bonnie, were born into this union. They lived on the farm for five years before their love of Sycamore Springs got the better of them. They purchased Sycamore from Noah Edelman in 1949 and operated the springs for 17 years (1966). Prior to that, my father owned and operated a Massey-Harris farm implement business at Sycamore and Hiawatha.

In 1966, they sold Sycamore Springs and moved to Sabetha where Chuck Dornes worked as a car salesman for Brey Motors, and Mary opened Mary D Dress Shop. Before retiring, Chuck sold real estate. (He only had an 8th-grade education. He studied and received his GED to qualify for real estate sales.) Following his retirement, Chuck enjoyed spending time in his woodworking shop and traveling.

> Charles Dornes died on April 22, 2007 at the Sabetha Community Hospital at the age of 89. He was preceded in death by: his parents; one brother, Maurice Dornes; and a granddaughter, Amy Jo Dornes. He is survived by his wife, Mary; and one sister, Leona Murchison Wikle of Sabetha. Other survivors include two children: Daniel Dornes of Sabetha; and one daughter, Bonita Hanni; and 6 grandchildren (David, Kevin, Melissa, Bradley Dornes and Jody Becker, and Sam Hanni).

After moving to Sabetha, Mary worked as a Dental Assistant for Dr. John Stone. In 1969 she opened the Mary D Dress Shop on Main Street. She owned the store for 17-18 years and served coffee every day at the store while the women shopped. Women came to sit on her sofa and talk about their day. A group of women started coming every day.

This group of friends became the "Hysterical Society." They antiqued

Mary Dornes 2019

together, played bridge, golfed, bowled together, laughed, and cried together. If you were looking for someone, you would come into the store and see if anyone had seen them.

Mary Dornes passed away in 2020 after a long fulfilling life.

Her motto to live by was, "Always look at both sides of the story."

2019

After the store was sold, Mary D started having "Monday Morning Coffee" at her house. This same group of women came to the house and were served coffee and a sweet on her best china and crystal. Again, when you were looking for someone, they would stop by and see if anyone had seen that person. The Hysterical Society, not the Historical Society, lasted for many years. They raised their families together and were true friends.

Dan and I received an allowance for helping with the happenings at Sycamore. Dan received 50 cents and I received 25 cents. We went to Sabetha on Saturday nights to spend our allowance. Dan bought Superman comic books for 5 cents from Darville Drug Store, and I would buy paper dolls from Duckwall's Variety Store and candy, or salted pistachios from Greene's Drug Store. Greene's made little vials of cinnamon oil for 10 cents too. Dip a toothpick in the oil and then suck on the toothpick flavor. My Dad always bought peanut clusters at Duckwall's. Mother would shop at Haines Clothing store or Geiger Jewelry store. My Dad sat on the hood of the car and talked to all the guys passing by. This was one night away from Sycamore Springs.

The skating rink and swimming pool were never open on Saturday nights. My Dad had promised to take our family to Sabetha on Saturday evenings. We would eat din-

Debbie Reynolds Paper Dolls

Bonne Toy Collection

ner at Buzz Café—they made the best scalloped chicken and homemade caramelized cinnamon rolls. Then we would do our shopping and see all of our friends.

When I was older and we went to Sabetha, this was the time to watch the boys drive by and yell at my friend Gayla Manche and me, or we would get to ride with them. Sabetha had one U-turn at the end of Main Street. The cars would make a U-turn there and then drive out to the highway and turn around at the Dairy Queen or the teen spot, Kid-o-lounge. It was one mile from there back to the U-turn on Main Street. One night I wanted to take the car. My dad checked the mileage before I left. I drove 28 miles that night just back and forth on Main Street. He was very conservative and gave me so much grief.

Superman Comic Book

Church Family

Sundays were important to our family. We attended Rock Creek Church of the Brethren. We didn't open the skating rink or swimming pool until 1 p.m. on Sundays. We would come home from church and smell the pot roast that had been cooking all morning. I still think of those times when I smell roast beef.

Rock Creek Church was a small church, small with a congregation. There were about 13 families that came every Sunday. We did all of our social gatherings together. The Lord was the center of our lives. Babies born, weddings performed, ice cream socials, pancake suppers, Mother and Daughter banquets, and our annual harvest festival were attended by all.

It was at Rock Creek Church that I gave my life to Jesus Christ in 1968. Charles Wells was our preacher at that time. His life had turned around and he shared his new faith in Christ with our congregation. My decision also changed my life forever. I attended Moody Bible Institute in Chicago because of faith. The foundations formed in my youth have given me direction all my life.

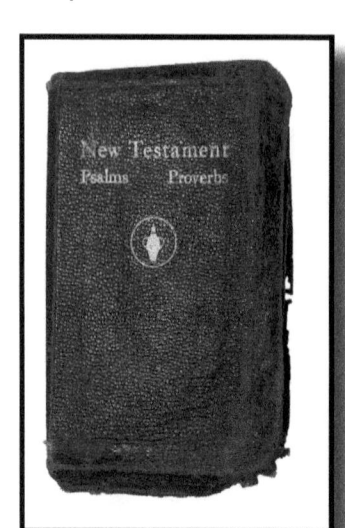

Gideon New Testament

It seemed like the church family all grew up together. We laughed, we learned, we celebrated, and we cried when one member died leaving a hole in our group, it was our family.

Winter Break

Wintertime at Sycamore brought change to how the resort operated. This was a quieter time for my family. It was a time of renewal. Dad would work on repairs that were needed. Mother concentrated on our home. My mother grew up listening to classical music. We had a stereo and there was always music playing on the 33-rpm vinyl records. Rodgers and Hammerstein music, Grand Canyon Suite, Madame Butterfly, Sound of Music, My Fair Lady, Barbara Streisand, Ray Coniff Christmas albums, and Smothers Brothers were some of our favorites. This gave me an appreciation for many types of music.

We would sled-ride down the hill by the house with our friends Shirley and Belinda Kendall. So cold, but so much fun. The neighborhood would get together to go bob-sledding and sled-riding.

Moody Bible Institute Banner

The Vern Moehlman family had an ideal hill to bobsled on. Our neighbors and church families would play together. There was food prepared by the women after everyone was exhausted from the sled riding and winter cold weather.

Gazebo Spring

Winter at the House

Winter on the Front Steps

Winter at the Skating Rink

Winter at the Pool

Winter on Pony Creek

Life outside of Sycamore Springs

Dan and I were involved in Morrill's school activities. Dan played basketball and baseball. We never had a football team at Morrill. Legend says that many years ago a boy was killed playing football so they stopped having a team. Baseball was popular and we played at area schools each year.

I also played basketball and softball. I was a cheerleader in Jr High and High School. Morrill had good teams and won many games. We played all of the small towns in the area: Reserve, Fairview, Hamlin, Willis, Seneca, Powhattan, and Robinson. Wintertime was busy with school activities.

Dan and I were involved in Morrill Tip-Top 4-H Club for 5 years. I learned how to cook biscuits, cakes, and bread and basic sewing (apron and hemming tea towels, and dresses from Mrs. Nathan (Lois) Grimm) while Dan raised Giant Checker rabbits. Dan had cages for the rabbits behind the house.

1966 Year Book, Morrill High School

4-H Pledge: I pledge my head to clearer thinking, my heart to greater loyalty, my hands to larger service, and my health to better living, for my club, my community, my country, and my world.

The skating rink was open in the wintertime on Friday and Sunday evenings. Dad ran the rink by himself on those cold nights. It was always surprising how many people came when there was deep snow everywhere. Dedication to having a good time with friends.

We had good neighbors—Danny and Margaret Kendall and their daughters Shirley (Brougher) and Belinda (Greiner). Danny and I played at their house.

4-H Pin

Danny & Margaret Kindall,
Shirley Kay & Belinda, 1950s

Danny & Margaret Kindall,
Shirley Kay & Belinda

Charles & Mary Dornes,
Danny & Bonnie, 1966

We loved playing cards or building straw bale forts and throwing snowballs in the John Nieman barn across the road. The Kendall girls had a corn crib that was their playhouse. (Our playhouse was in one of the cabins north of our house.) We would walk across the pasture between them and Sycamore to play together. We had to be on the watch for the bull that was in the pasture.

Danny Kendall mowed our pasture area for one of our formerly hired girls, Susie Mishler, so that her husband could land his small plane near the Springs. What a treat for our families to fly for the first time!

In the winter when we had snowstorms Danny would drive his tractor down to Sycamore to pick up our family and take us back to his house so we could have a fun evening playing cards. Amazing what we did to be with our friends.

Aerial View of Swimming Pool

At the Pool, Ray Harris, Gayla Manche, Bonnie Dornes

Danny Dornes on the Bike

Bonnie Dornes Age 7

Chapter 13
Today

Dan, Mary, Bonnie 2018

My parents are gone now but my brother, Dan, and I still talk of our times at Sycamore Springs.

(Dan Dornes, Mary Dornes, and Bonnie Dornes Hanni visited Sycamore Springs in 2018.)

This is an updated picture of the large swimming pool, but it looked much like this in the 1950s and 1960s. It held 500,000 gallons of water, one of the largest pools in the area.

The swimming pool opened every year on Memorial Day and closed on Labor Day. It took 10 days to fill the pool with natural spring water.

The pool held ½ million gallons of water that came from the spring across the creek. Dad built the diving boards by gluing 2x4 wood together and sealing them with shellac.

Dad didn't know how to swim but he cleaned the pool every morning at 5:30 by pushing a large broom along the bottom to move the dirt to the deep end. Then Dad, and usu-

Swimming Pool Ready for Summer

120 | Bonnie Dornes Hanni

Swimming Pool Towards the Deep End

ally a lifeguard, would drag the deep end with a large vacuum to remove any debris. HTH Chlorine was used to sanitize the water.

Our first Lifeguard was Don Wenger from Sabetha. He was one of the best-looking guards. We had many lifeguards over the years until Dan was old enough to be the lifeguard, and I later served as the guard for a number of years too.

The pool was used in the morning for swim lessons that ended at noon. Mom and Dad, and all of our workers, would take a break and go over to the house for lunch. The pool was then opened 1 to 10 p.m. The cost was 50 cents each day, or we offered season tickets for those families who came every day.

During our lunch hour, the guys played horseshoes. They would get down on the ground to see whose horseshoe was closest to the pin and who won. In later years, we watched Amos and Andy on TV. Then it was time to go back to work.

There were several years when we had go-karts and a track out by the ball diamond. Guys would bring their go-kart and there were many races. Dan often took his kart to Morrill where they had a go-kart track too. Charlie Wissler from Sabetha was one of the kids who also participated too.

As little kids, Danny and I were in and out of the pool all day. I had more swimsuits than summer clothes. I wore out the suit seat from sliding off

1950s Skating Rink & Swimming Pool

the cement side of the pool or sitting on the side of the pool. With my blonde hair, I was called "Bonnie Braids"; it would turn chartreuse green on the ends of my braids from the chlorine that was put in the pool.

Dan and I were only allowed one treat a day from the concession stand. Everyone always thought we had it made but we had rules to follow. We had a pretty normal family life. We caused trouble and had to be punished like other kids.

Some of our lifeguards were Ray Harris, Jim Mitchum, Fred Fulton, and Jenny Fulton. One year Fred Fulton decided to jump from the high dive down to the low dive and then dive off from there. The diving board broke. What did Dad do? He built a new one and moved on.

I am sure there were other lifeguards. When we got older, Dan and I were lifeguards. We had the best tans. Never wore sunscreen. Sometimes I would use baby oil. Imagine what effect that had on my skin from the sun. It was a big responsibility to watch the pool for anyone struggling to swim. We never had any fatalities.

Swimming Show Friday Ends Program

Swimming Show Program, 1961, Sue Nesbitt, Charlotte Welck, Betty White

Mother served the clothes baskets to the swimmers in the rack room. She would see little kids in trouble and quickly climb over the counter and jump in the water to save a struggling kid. There was always some kid who ventured too far to the deeper water.

Some summers when the pool officially closed on Labor Day, my Dad was approached by a group of Nuns from a town in Kansas to see if they could come to the pool for the day. We weren't allowed to be at the pool beside Dad with the Nuns. They came for several years.

Mary Dornes in the Rack Room

Bonnie Dornes by the Pool

Swimming Events

The show "Cavalcade of Old-Fashioned Swimmers" was the special feature, Miss Jenny Fulton, lead instructor. The old-time suits worn by the girls were sent by request from the Jantzen and Catalina swimsuit manufacturers on the west coast.

The Sabetha girls were teachers with the summer swimming program. Every year at the end of the summer there was a Swimming Program. The show began at 7 p.m. and all of the students participated by performing what they had learned — going underwater, floating face down, back float, and swimming on their stomachs, diving, breaststroke, and backstroke. The teachers ended the program with a synchronized swimming performance that was organized by Jenny Fulton. I was a part of the performance many summers.

Swim lessons were offered every summer. There were lessons for beginning swimmers and continued Life Saving classes. Buses would pick up the children at the Sabetha library park and drive 7 miles to Sycamore. Each lesson cost 10 cents. The classes lasted an hour or until the bus came back with the next advanced class of kids. There were beginners, intermediates, swimmers, and life-savers classes. It was always a treat to buy fresh popcorn, a pop, a candy bar, an ice cream cone, or candy at the pool snack bar.

The program at the end of the summer swimming lessons (as mentioned above) was a highlight for each swimmer. Cars would park along the fence and parents and friends would stand outside of the fence and watch their children perform what they learned. It was quite an achievement for the little kids to be able to go under the water and not be scared.

Since Dan and I were always at the pool, we passed each class with flying colors. I think I was 12 and had to wait until I was 16 to be in the Life-Saving class.

In the life-saving class, it was important to learn how to save someone who could not swim, especially in the deep end (10 ft). We

Life-saving Badges

had a heavy-weight lead slab that originally was from the suit that was used to go to the bottom of the pool and sweep and clean the pool. This slab was thrown to the bottom of the pool and each person had to retrieve it. It was very heavy as I remember. We also had to save one of the swimming teachers, and Fred Fulton was one who put up a fight to be saved. Very scary to try to save him. I passed.

Father Neptune

The final event was the appearance of "Father Neptune." This character was played by various lifeguards and eventually by my brother Danny, performing this role. He would climb to the top diving board and stand at the end and speak to all of the children. He would praise the children for all of their efforts in each of the lessons. He would tell them to be good children and to obey their parents. He would be watching when they come back next year to learn lessons in how to swim.

Father Neptune was dressed in a cape (a towel around his back), goggles, and flippers on his feet. He told the children that he would see them next year and proceeded to dive off of the high dive. He would swim underwater up to the shallow end, and come up amid a group of teachers and climb out of the water.

The Kids thought Father Neptune lived under the water and watched them during their classes. Father Neptune was a great tradition for the summer swimming program.

Every Spring when the rains came, we had to worry about flooding. In the 1950s there were no conservation rules for the land. Eventually, government-mandated ponds and dams were built to control the overflow of water on the land.

Kids by the Pool

Pony Creek was a small creek but when we had heavy rains the creek would fill and overflow around the buildings. Our house was about 50 feet from the creek and was much lower than the creek. Water would overflow and come into the house. The house had a café and kitchen in the basement and we always had a foot of water come in on the floors.

I remember one summer that we had camp kids staying in the hotel and camp house, Camp Kanebwa. The flood left a foot of water in the café where meals were served to the kids. Camp counselors and our hired girls and Mom and Dad would scoop the dirty smelly creek water out of the café to get the room ready for the meals. Mom was frantically working in the kitchen preparing the meal and everyone else was wading through the water to get it out of the house. During one flood, I could reach out of the café window and touch the flood water. It was that deep around the house.

Every October was a special time at the skating rink. Decorating the rink was so much fun for the annual Halloween party. My Mother loved Halloween.

We used tree branches with all of the Fall colors. We drove our 1954 Blue Chevrolet pickup up to a place we called "Lover's

Lane" close to my Grandparents' home near Salem, Nebraska. My Dad would cut large oak limbs with fall colors, red sumac along the road, and wildflowers, like bittersweet. The pickup was filled with many colorful limbs.

We bought our first car in 1956. Chevrolet all the way!

Early days of Halloween decorating, driving up to Salem to get the fall leaves in 1950. The author with her kerchief. Danny with his hand over his eyes.

Flood by the Shop

Mother had old orange papier mâché pumpkin lanterns that were put over the lights in the rink. There were also black cat lanterns we made out of construction paper with the eyes open to let in light. It made it dark in the rink and this made it much easier for the couples to neck while skating.

A lady from our church, Eva Aldefer, made homemade doughnuts. Eva would make 13 dozen glazed doughnuts. The doughnuts along with apple cider were free to all of our customers and skaters. Many people came to enjoy the treats.

Flood by the House

So many kids came from Nebraska to skate. Humboldt, Dawson, Shubert, Falls City, Barada, Salem, Table Rock, and all of the other little towns north of Sycamore came to the rink to skate. Many skaters also came from the neighboring Kansas towns such as Morrill, Hamlin, Reserve, Sabetha, Fairview, Seneca, Hiawatha, Soldier, Goff, Circleville, Holton, and many other small towns.

There was a contest between Kansas and Nebraska kids. Tug-of-war with

Flood of Pony Creek

**Kids in Pickup;
Bonnie in Kerchief**

a huge rope (the rope from the pool that separated the shallow end from the deep end) was the game. Being on skates made it hard to hold the line but someone had to win and the defeated team would fall.

Another game was "dress-up." In two teams, each person put on a special piece of clothing such as a shirt, an apron, a hat, or belt and raced to the end of the rink and back, taking off the garment and passing it to the next person.

We played Limbo on skates. Every person had to skate under the stick. Each time, the stick was lowered until the last skater, who could go under the stick, won. My brother Dan was often the winner.

So much fun with all who came for the evening.

Paper Mache Pumpkin

Open Skating Fridays and Sundays

Dan and I had our own skates and Dan still has his skates today. Dad was in charge of handing out the skates and Mother was in charge of the concession stand.

Mother made fresh popcorn every night. The pop was in a cooler that contained water to cool the glass bottles. We served only Coke, 76, orange, and grape flavors.

Chuck was the referee on the skate floor. He blew his whistle to make the kids slow down. Sometimes if they wouldn't obey, he would kick them off the floor, and if they still were causing trouble, he would ban them from coming back to Sycamore for 2 weeks. The funny thing, these guys respected Dad, and

Coke Sign

they always come back so they could skate once again.

We played many of the popular 45 rpm records for the skaters. Mother would go to Falls City to one of the stores and buy new records every two weeks. Elvis Presley, "It's Now or Never"; Chubby Checker, "The Twist"; Ricky Nelson; Roy Orbison; Johnny Cash; Beatles; Everly Brothers, "Don't Want Your Love"; Johnny Horton, "Sink the Bismark"; The Drifters; The Ventures; Beach Boys; Frankie Avalon; Paul Anka; Connie Stevens; Brian Hyland, "Itsy, Bitsy, Teeny Weenie Yellow Polka Dot Bikini"; Connie Francis; Marty Robbins, "El Paso"; Bobby Rydell; Brenda Lee, "I'm Sorry"; Percy Faith; Summer Place; Hang on Sloopy; Tom Jones, "Pussy Cat"; Barbara Mason, "Are You Ready"; The Byrds, "Tambourine Man"; Mamas and Papas; Herman Hermits; The Righteous Brothers, "Unchained Melody"; "Teen Angel"; "Running Bear"; "Moonlight Swim"; "Personality"; and many more rock and roll songs.

Coke Water Cooler

There were different times of fun on the skate floor. We had couple skates and triple skates. The guys would choose a girl partner, skate together and when the whistle blew, the guy moved up to the next girl. The triple skates were usually with 2 guys and a girl. When the whistle was blown, the guys would move up to the next girl and proceed to skate until the whistle was blown again. This was one way for everyone to meet new people.

In Mom and Dad's Day, the time at the rink was a way to meet a new girlfriend or a boyfriend. Some even became serious and got married.

I was a pretty good skater. I could skate sideways, and twirl in circles. My brother Dan was a much better skater. He could do all kinds of tricks like squatting down and skating on one leg. We liked twirling together.

Record Player

There was no air conditioning in the rink. There were windows all around and they were left open to let in air. It got really hot in the rink with so many skaters on the floor, cold in the wintertime too.

Louise and Chet Bloom with daughter Nancy (Brockhoff) came one summer and lived in one of the cabins while he painted the skating rink. There were always repairs to keep the place looking appealing.

For several years on certain Saturday nights, we closed the skating rink only for the African-American kids. Dad was the only white guy in the rink. People from as far away as Atchison came for a night of skating.

I remember that not too many Blacks came to the resort. I do remember one guy who came and taught me how to do the Twist. Chubby Checker's, "The Twist " was a favorite song.

Danny Dornes 1955

One of the funny things (not really funny, but...) Danny did when he was a little kid, probably 7 or 8, was trying to help Dad. He decided to rake the leaves under the rink, but there were so many that he decided to burn the leaves. Dad saw the smoke and was able to put out the fire before the rink burned down. Danny got in trouble. He had to stay on the other side of the bridge for a week and stay away from the pool and rink. He and I made a game out of it and so it wasn't like he was punished. We explored the woods during this time.

When the Rink closed every night at 10 o'clock in the summer, we would take off our skates and soak our feet in the swimming pool. Sometimes we would end up swimming in our clothes.

On Friday nights after the rink closed, we would pile into a car or pickup and drive to Sabetha to get something to eat. In the early years, we went to Nabb Meyer Café, known as Nabb's. We sat up at the counter on the stools and made the best hamburgers.

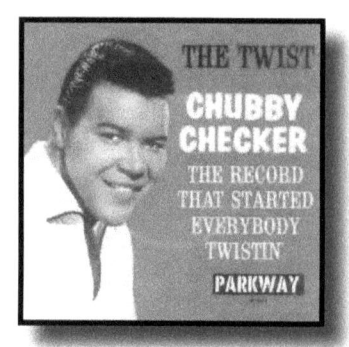

Chubby Checkers, The Twist

The guys would play pinball. It got pretty competitive and rowdy.

One summer Dad decided to learn about refrigeration repair. We never quite figured out

why he wanted to do that. His enthusiasm didn't seem to last too long. But he had purchased a home course on refrigeration, and he decided to sell the course. A man from Hiawatha offered to buy it. Dad and I drove over to Hiawatha to deliver the course. The man paid us by giving us an older Vauxhall car. It was a tiny lime green foreign car, a funny car for us. On those nights we went to Sabetha for a meal, we piled into the Vauxhall and be on our way. We could have made the Guinness World Records book for how many people could fit in the car. It was a laughable experience to own that type of car. Dan drove it as his own car.

Bonnie & Danny with Badminton

A Vauxhall car

Danny Dornes First Day of School

After Nabb's Station went out of business, we started going to Midway Café. Flossie and Kurt Wertenberger owned the gas station next door. They became good family friends. The café was small, so when we arrived, it became overflowing. Many times, some of the skaters would come to town with us.

We would eat breakfast and sit around the same table. Jerry and Gene Niehues rode their motorcycles to town and joined us. Walt Arnold and Ronnie Gibson from Falls City, Nebraska, who

Danny Dornes Age 9

helped my Dad at the rink, became good friends and would come to Sabetha with us before going back to Falls City. So many good friends were made. We laughed a lot.

One of Dad's jokes was taking care of the imaginary "anteater" as a way to keep the laughter going. Good times, memorable times. We always had a traditional breakfast at 10:30.

Camp Kids—Camp Kanebwa

When Mom and Dad started their new venture of owning Sycamore Springs, a church camp group was coming to stay one week at a time for 3 weeks. Each week there were 100 kids. They would stay until Saturday and then the next 100 kids would arrive on Sunday.

The first week we had camp kids, we had a flood in Pony Creek. Water came into the cafe with up to a foot of mud and water. The camp counselors hired girls, lifeguards, and anyone else we could ask to help clear out the water and mud to get ready to serve the meals. I can still smell flood water and the smell takes me back to those times. Meals were prepared in the lower-level, summer kitchen and served in the café.

Vauxhall Car

The camp was called "Camp Kanebwa." Jr. High church kids from Kansas, Nebraska, and Iowa would travel to Sycamore for a fun time and a time of faith learning.

Camp Counselors would stay for 3 weeks. Our family became their family and we all worked together to make the experience for the kids memorable. Being young, Danny and I were able to take part in all of the camp activities.

The girls stayed in the hotel while the boys stayed in the camp/shelter house behind the skating rink. Army cots were lined up in the camp house for all of the boys and counselors.

Mom hired 2 or 3 high school or older girls—Susie Mishler, Georgia Windle, Rosie Edelman, Jane Isch, and Julia Feek, to name a few—to help with the camp. The girls became big sisters to me. They also had to keep an eye on Danny and me because Mother was so busy. One

time they couldn't find me; I was pretty little, probably 3 or 4. Everything came to a halt. Everyone was looking for me... in the creek, in the hotel, in the woods, nowhere to be found. I had gone to my bedroom and crawled into my doll bed to take a nap, covering myself with a big old comforter. After they found me, it was time for everyone to get back to work.

In the summer we moved our kitchen from the main floor of the house to the basement kitchen. No air conditioning so it was much cooler in the basement.

Breakfast, lunch, and dinner were served to the camp kids every day. As a little kid, I helped stir the morning hot chocolate in very large pots. If I was slacking at my job, the scum would form over the top of the cocoa and I had to be diligent in doing my job. I also washed the silverware—100 forks, 100 spoons, and 100 knives for each meal, each day. Georgia Windle was in charge of making the cinnamon toast with 13 loaves of bread.

The camp had many activities. They would skate and swim, but there were other lessons to be learned. Small groups venture into the wooded area behind the house and have a nature lesson, focusing on how God created the earth and all that is on it. They made terrarium in gallon glass jars with the moss, and small plants they found in the woods.

Vesper services were also held in the woods. The counselors had a portable organ that they would haul up into the woods behind the house and sing hymns or fun camp songs. Many chigger-bites were had by all.

Campfires were built on the ground with rocks in a circle, and a meal was prepared over the fire. Mother would make aluminum foil packets with hamburgers, potatoes, and carrots that were put in the hot coals to cook. While the kids were waiting for them to cook, "walking salads" were made with cabbage leaves smudged with peanut butter all over and then folded, and then campers could walk around while eating the salad. We would make smores to end the meal.

Every evening the group would gather on the patio in front of the hotel. Kids were sitting on the ledges or in the grass. Camp songs and a Bible lesson were given to relax the children to get them ready for the night.

During these 3 weeks, Danny and I became a part of the group. We had other responsibilities, but we did get to make all of the craft projects offered. Back to the evening gathering... Dan and I would sneak up to the roof of the hotel and peek over the edge to watch the kids sing and hear the lesson. We took small rock pebbles and dropped them over the edge onto the kids. Then we would hide. I am sure the camp counselors knew we were doing this, but we had so much fun doing it and we had to be very quiet and not laugh out loud.

We learned how to make neck lanyards out of different colored plastic-covered cording. We also covered metal hangers with this colorful cording. Bracelets and rings, even earrings were made with this cording. The hangers were used for many years. I learned how to do all of this easily so I became a teacher-helper for those needing assistance.

The lanyard I made in the 1950s. A name tag was attached at the bottom.

Maybe this was the beginning of my art endeavor that has carried me through all of these years. I have learned from my mother how to do every craft project such as knitting, crocheting, needlepoint, embroidery, and paper projects, and now I have a college degree from the University of Kansas in textiles, weaving and fine arts (2002).

I received my master's degree at Colorado State, Ft. Collins in textiles and went to Chicago to be a professor at the Illinois Institute of Art teaching historical fashion and textile technology (2006). I still love working with my hands on craft projects.

Back to Camp Kanebwa.

When I was little there would always be a camper who would get homesick. I became their little sister, and they would play with me and my dolls. I dressed up our cats too so there was always a way to satisfy the sad camper.

At the end of each week, the hotel and camp house needed to be cleaned. Back in those days, sheets were 100% cotton and wrinkled easily. Spring water is highly mineralized; we called it "hard water." It was difficult keeping the sheets and towels white from the rusty water.

Mom used the "mangle iron" to press all the sheets to get them ready for the next week. Saturdays were very busy for the hired girls and especially Mother. Buying food for the meals was quite a task too. We never had a garden; there was no time for that, so the grocery store was our garden.

Lanyards

Hard work, and hard times, but we made it work. Mom and Dad wanted everyone to feel at home. This is the time I learned about being of service to the public.

Camp Kanebwa lasted for a number of years until the various churches started their own camp sessions. So many people we met along the way. It was a great time in my life to be with so many kids and also to learn the craft projects and sing the songs.

Mangle Iron

Were you one of the campers who came to Sycamore Springs for the summer?

Jerry and Terry Tietjens, Owners
Sycamore Springs 1966 1984

The twins were entrepreneurs of many historical venues. The Springs were sold to Jerry and Terry Tietjens in 1966. The Tietjen twins added many improvements that are still seen today such as miniature golf (1973), lowering the skating rink ceiling, synchronized disco lights installed, and remodeling the concession area.

Other improvements were: The spring gazebo was built to spotlight the running crystal-clear water. A pump house was built to analyze the properties of the water. A sales office and general store for the campers were built as you enter Sycamore Springs. Cement block restrooms for campers and shelter houses for picnics and camping were improved. At this time, people started bringing their campers to enjoy the activities at Sycamore Springs.

Terry & Jerry Tietjens

Camp Store & Antique Store

Skating Rink with
Disco Lights

Spring Gazebo

Spring Gazebo on
the Hill

Water Chemical House
in Winter

Trailer camping became popular

In 1969 trailer camping became popular and the first hook-ups were installed. In 1974, Dennis Dodson, Jr. joined the firm. An antique shop was built with an area for camp and picnic supplies.

Rock Creek Church of the Brethren, later called Rock Creek Chapel, was originally located on Pennsylvania Avenue north and west about 5 miles from Sycamore Springs. With the closing of the church in the 1970s, Terry and Jerry Tietjens arranged for the church to be moved to its current location in 1979.

Camper Area

Promotional Brochure

Since that time, the Tietjens added the eight-rank pipe organ, originally from the First Baptist Church of Hiawatha, KS. The organ was completely renovated. Terry often played the organ for entertainment.

Services were held in the Rock Creek Chapel every Sunday by visiting churches from the vicinity.

Square Dance Brochure

The Tietjen twins were starting a historical frontier village main street as an added attraction at Sycamore Springs along with the Morrill Railroad Depot and jail.

The Gatekeepers of Sycamore Springs | 137

Morrill Train Depot now located at Sycamore Springs

The Saint Joseph-Grand Island Railroad was completed in 1870 before the town of Morrill Kansas was incorporated in 1886. The first depot was a freight car before the building was erected in 1877.

Morrill Train Depot

The Depot was moved to Sycamore Springs to preserve its historic significance. The Tietjen twins, Jerry and Terry, painted and renovated the building to showcase its history.

The Depot is now the Museum for Sycamore Springs Whitetail Ranch. Photographs and physical articles can be seen from past memorable days. If you have items relevant to Sycamore Springs, please donate them to the museum.

The Tietjens moved to another historical area, Abilene, Kansas. They maintained the Seelye Mansion, an "8 wonders of Kansas-Architecture" and gave tours to explain the importance of Seelye in creating patent medicine in the area.

Rock Creek Church

Church Organ

Unfortunately, Jerry passed away, but Terry still manages and maintains the mansion, a historical treasure.

THE KANSAS STATE DEPARTMENT OF HEALTH
DIVISION OF LABORATORIES
Environmental Health Laboratory
801 Harrison Street   66612

TOPEKA
KANSAS

E. D. LYMAN, M. D., M. P. H.
Director of Health

November 2, 1970

Mr. Terry Tietjens
Sycamore Springs
Route 2
Sabetha, Kansas 66534

Dear Mr. Tietjens:

    Listed below in milligrams per liter are the results of a partial chemical analyses of two samples of water collected October 19, 1970, from your springs known as the Sycamore Springs.

	Spring No. 1		Spring No. 2	
Total Hardness (as $CaCO_3$)	= 1760.	mg/l	1750.	mg/l
Calcium (as Ca)	= 628.	mg/l	620.	mg/l
Magnesium (as Mg)	= 47.	mg/l	49.	mg/l
Sodium	= 51.	mg/l	40.	mg/l
Total Alkalinity (as $CaCO_3$)	= 298.	mg/l	286.	mg/l
Chloride	= 78.	mg/l	61.	mg/l
Sulfate	= 1400.	mg/l	1400.	mg/l
Nitrate (as NO_3)	= 5.3	mg/l	6.2	mg/l
Fluoride	= 0.8	mg/l	0.7	mg/l
Iron	= 0.15	mg/l	0.31	mg/l
Manganese	= 0.00	mg/l	0.00	mg/l

    These waters are very hard and highley mineralized with sulfate. Such water will have a strong mineral taste and it will produce a laxative effect on people who are not accustomed to drinking it.

Sincerely,

Nicholas D. Duffett, Ph.D.
Director

Howard A. Stoltenberg, M.A.
Chief, Water Chemistry Section

HAS:glb

Chemical Properties of Spring Water

Chapter 14
Sycamore Springs Community

Rock Creek Church on Hwy 75

Rock Creek
Church of the Brethren History

During the western expansion years (1878-1886), several Church of the Brethren members from Illinois, Pennsylvania, Virginia, and Iowa moved into Brown and Nemaha Counties of Kansas. By 1881 about 200 members were living in the Pony Creek area north of Morrill, Kansas. Services were held in the Rock Creek School until the church was built in 1888.

The church was the center of the community. When there was a need or a time of fellowship and celebration the neighbors came together.

NEIGHBORS HELPING NEIGHBORS IS THE SUBJECT OF THIS WEEK'S PHOTO FROM YESTERYEAR. When Forrest Payne was in the hospital in the 1950s, his neighbors pitched in to saw wood at his farm in the Rock Creek neighborhood. Helping were Daryl Bechtelheimer, Roy Holbert, Donald Windle (partially hidden), Harold Kesler (partially hidden), Langdon Livengood, Bob Schmidt, Louis Manche, Maurice Argabright, Gene Kellenberger, Orville Bailey, Harold Somers, Floyd Mishler, Leo Livengood and Harry Moore. Photo courtesy Langdon Livengood

Farmer Community

Neighbors in the Church area. Do you recognize anyone?

They were cutting wood for a neighbor in need.

Annual Ice Cream Socials were held with homemade ice cream, pies, and cakes. The church family made ice cream with the old crank ice cream freezers. The kids added the salt to the freezers to make the ice colder and the ice cream freeze faster. After we had made the ice cream, vanilla, chocolate, and butter-brickle made with chopped Butterfinger candy bars, we made a couple of gallons extra so we could feast after all the hard work.

For Pancake Feeds, the church men fried eggs, cooked the sausage or flipped pancakes and added hot syrup while the women served. This was a community effort and many people looked forward to this time in the church.

Harvest Festival was a time to celebrate farmers' crops and be thankful for the bounty. The altar was decorated with pumpkins, squash, variegated-colored Indian corn, and fall leaves.

Vacation Bible School was a yearly time for the children. Thelma Funderburgh was the main teacher. Thelma lived near Sycamore so she would pick us up in her car to drive to the church. She loved going over the hills singing "Down in the meadow was an itty-bitty pool, swam 3 little fishies and a mama fishie too. Swim as fast as you can. Swim if you can all over the dam." We would go fast over the rolling hills singing and laughing all the way to the church.

Ester Orr played the organ and Bertha Manche played the piano. (She always liked the song "Sweet Hour of Prayer" for our church services.) At Christmas, we would get to hear Marvin Hinton play the organ and sing "O Holy Night." With the candles glowing in the windows and the lights turned low, we were all feeling the Christmas story come to life.

Faye Bechtelhiemer had a beautiful flower garden. She supplied the flowers for the table in front of the altar every Sunday. I can still smell the purple iris and the colorful peonies that were in the bouquets.

At the end of Vacation Bible Study (VBS) week, we had a program. We stood on the stage to sing and say our Bible verses. I memorized Psalms 100, and I can still say it today. Memorization of Bible scripture influences young minds. It is amazing how God has used that psalm in my life.

These activities were a part of the ongoing activities that helped the church flourish.

> "Make a joyful noise unto the Lord, all ye lands.
>
>> Serve the Lord with gladness: come before his presence with singing.
>>
>> Know ye that the Lord he is God: it is he that hath made us, and not we ourselves; we are his people and the sheep of his pasture.
>>
>> Enter into his gates with thanksgiving, and into his courts with praise: be thankful unto him, and bless his name.
>>
>> For the Lord is good; his mercy is everlasting, and his truth endureth to all generations." —Psalms 100

Rock Creek Church of the Brethren was located on Pennsylvania Avenue North of Sabetha, Kansas on old Highway 75.

When the church closed, a few families met in members' homes with Faye Bechtelheimer as the teacher/leader. It was a small group but dedicated.

This was the church moved to Sycamore Springs by Jerry and Terry Tietjens. Many weddings were performed in the small country church. My family were members and attended there all my life. Seeing that the church building was preserved at Sycamore Springs was wonderful and fulfilled a need for those coming to the Springs.

The Church building was moved to Albany in 2023.

This little handmade, hand-painted cross was given to me as a youth by Jim Kesler who attended Rock Creek Church at the same time as I did.

Somerset Pennsylvania

The Revolution of 1848—49 in Germany was a pivotal point of immigra-

Yellow Cross

tion to America. The country was going from feudalism to capitalism to socialism. The first German immigrants came to the United States seeking land and the promise of religious freedom. Descendants of the first German immigrants are called "Pennsylvania Dutch," an Anglicization of the word "Deutsche" meaning German. They had heard that both could be

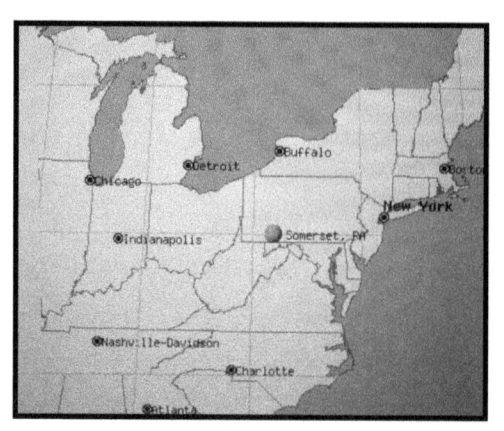

Sumerset Pennsylvania Map

found in the new charter colony of Pennsylvania, which was governed by a Quaker, William Penn.

In 1854 Congress created the New England Emigrant Aid Society after the Kansas—Nebraska Act was approved. The Kansas Immigration Society established in 1871 brought more people to this rugged land. Railroads were being built to bring people to the West.

Somerset, Pennsylvania was home to many of the settlers who traveled to Kansas and settled on Pennsylvania Avenue north of Sabetha. Many of the settlers didn't come directly from Pennsylvania. They often moved west to Ohio, Illinois, Minnesota, and Missouri, and stayed in those areas while others would move on to Kansas.

The land has gently rolling hills, and there is a lot of bottomland along the creeks on both sides of the avenue. The flat land was the prime land to own. The land reminded them of their homes in Germany.

This 1983 Stamp gives recognition of 300 years of German

immigration to America.

In the late 1860s, pioneer and acting land agent, Jonathan Lichty, came to Kansas in search of farmland for his family. The family had been waiting in Pennsylvania until a large parcel of land was purchased on behalf of the group. He also financially helped other settlers to purchase land in the Pennsylvania Avenue area.

German Immigration First Cover-Stamped Envelope

The settlers that followed in 1870 were: John Lanning, Jacob Myers, Jacob Heikes, Elias Saylor, Allen Lichty (Vernelle Mock's grandfather), Peter Faidley, D. Spangler, William Flickinger, Bitner, Peter Brougher, James Livengood, Noah M. Kimmel, Aaron Kregar, Jonathan Scott, Beachy, Caroline Wolford, Phillipi, Peter Lanning, Uriah Saylor, C.Babst, Dan Saylor, C.A. Saylor, John Hoover, A.M. Schaulis and John Fletcher.

Do you see any familiar family names?

The great-grandparents of Donna Phillipe Leman came from Pennsylvania in 1872 and settled one mile north of Sycamore Springs. They built their large barn in 1879. It was a very large fancy barn as I remember. The barn was built with wood pegs, no nails. We called it the Nieman Barn because the Nieman family lived there at the time. We had many snowball fights and built forts out of straw in the winter with Shirley and Belinda Kendall, and Terry Olberding.

Having come this far together, the friends chose to make their homes together along a road that ran east and west. Today it is known as 330th St., one mile south of the Nebraska border. They wanted a rich life to raise their families and something of that old life they had left behind.

The settlers built small homes initially near the Pennsylvania Avenue area but as their families added more children, they began building bigger houses. A competition between the size of the family and the size of the house ensued. As many as 80 or 90 children would attend one of the schools.

Several one-room, one-teacher country schools, and churches were built along Pennsylvania Avenue. The Avenue extended from north of Morrill, Kansas, and west to old Highway 75 north of Sabetha originally and then extended on the west to north of Bern, Kansas. Church and school were the community activities outside of farming.

Livengood Homestead

The church services were held in homes or schoolhouses. The Dunkard Brethren Church was the first one built north of Morrill. In the 1880s there was division in the Dunkard church in the United States, the two groups were called Conservatives and the other Progressives. "The Church of the Brethren" was conservative and "The Brethren Church" was more progressive. The Lutherans also organized a congregation in the 1870s.

Rock Creek Church of the Brethren was conservative. I attended this church all of my young life.

Chautauqua Speaker

The Salem, Nebraska Chautauqua was a Sunday School Convention Camp Meeting outreach to people 15—20 miles away and many people in the Morrill area attended the meeting on the banks of the Nemaha River. All of the Sunday Schools were invited to attend. They drove their farm wagons to Salem to attend the events. It was a vacation from normal farm life. There were tents to rent if they wanted to stay for the seven-day event.

Trains delivered people from other areas. Railroads reduced rates for those living in Nebraska, Iowa, Kansas, and Missouri so more people could come to the convention. Stables were available to care for the horses.

The town of Salem offered food and water for the visitors. Shopping was available in the town. Mail could be delivered in care of the Chautauqua.

A large tent that held up to 3000 people was where the convention was centered. Smaller tents offered Bible studies or other classes.

A program of speeches, lectures, singing, and concert bands came from many areas. The organizers brought entertainment and culture for the whole community with speakers, teachers, musicians, showmen, and preachers. It was not unusual for crowds of 5,000 to attend.

Chautauqua Speaker

Grandparents, Laurel and Martha Slayton lived near Salem and also attended these same Chautauqua events in the early 1900s.

Initially, the Chautauqua meeting was very religious in form, but eventually it became non- denominational and non-political. The Chautauqua was an event with informative constitutional scholarly communication as well as religious events with other parts of the United States.

Jonathan Lichty, who lived on Pennsylvania Avenue, was one of the organizing board members of the Chautauqua for this area. The Chautauqua movement originated in Chautauqua, New York in 1878.

The Avenue extended from north of Morrill, Kansas, and west to old Highway 75 north of Sabetha originally and then extended on the west to north of Bern, Kansas.

Sycamore Country Grade School

The Pleasant Valley District school is located ½ mile north of Sycamore Springs. It was open until the early 1950s and originally set on the hill south of Sycamore Springs by the upper gate. The school building was later moved to its current location.

The country school enrolled the neighborhood children. Nila Edelman Snyder attended school here and went to Sabetha for High School. Many community events happened at the school. The building was later turned into a hog house and was torn down in 2023.

Bertha Heikes Manche was the daughter of Robert and Rosa T. Rees-Heikes. Bertha was married to Harry Manche. She lived on the farm on Pennsylvania Avenue and taught school on Pennsylvania Avenue. Unfortunately, Harry passed away at a young age and Bertha became a single parent to Gayla, her daughter. They attended Rock Creek Church. Bertha Heikes Manche had been a teacher for many years, but the school closed the year that Danny was to start school.

Do you see any familiar faces? Class of 1938—1939. My Aunt Leona Dornes-Murchison-Wikle is in the back row.

Bertha Heike, Teacher

1938 - 1939 School Kids

1938 - 1939 Kids Names

1939 - 1940 School Kids

Bertha Heikes Manche

1939 - 1940 Kids Names

Morrill Kansas

I have included so much of Morrill history because it was my hometown, and where my brother Dan and I attended school. Dan graduated in 1964 and I graduated in 1967. Pennsylvania Avenue in Morrill is also closely related to events and people in Morrill.

Kansas was organized as a territory on May 30, 1854, and became the 34th US state in January 1861, while James Buchanan served as President of the United States.

Brown County was officially organized in the Kansas Territory in 1855. It was spelled "Brown(e)" until 1875 Secretary of the

Daughter Gayla Manche, Bonnie's Best Friend

State Board of Agriculture was named in honor of Senator Albert G. Brown of Mississippi, who served at the time of the passing of the Kansas—Nebraska bill.

The Middleswart family of Racine, Ohio staked claims to 160 acres in Kickapoo, located where the town of Morrill would be founded. They bought the land from the widow of George H. Green who was killed in the War of 1812. She had received it from the government for her husband's service in the war.

A Civil War soldier was granted the Kansas property (the government granted land to veterans who served in the Civil War) where Morrill was later to become a town. The soldier died while serving in the war. His wife sold the land to Thornton J. Elliot, also a veteran, for $2000 in 1867. He saw the sod broken in this new territory.

The town of Morrill had several buildings before it became a precinct. Elliott owned the property where the town was formed in May 1867. He was a staunch Republican and in 1888 was elected as Representative from Brown County to the Kansas State Legislature.

He gave much of the land to form the town. He also gave land for the cemetery after his wife Roxie and daughter died; built a Methodist Church; and a block for the new school. At that time, there were no other settlements in the area.

A Sac and Fox Indian trail ran west of the town of Morrill which was known then as Kickapoo, KS (N.E. section of the county). The Indians traveled from Hiawatha Kansas to Humboldt, Nebraska on this trail. They passed through settlements owned by J.B. McKim, William Sanner, R.L. Willard, and Jonas Forney.

There were two mounds of James Lane's old forts still visible near Sycamore Springs.

As the pioneers were crossing the Kansas area, they were shooting the buffalo for sport as well as a meat supply. This was taking away from the Indian's supplies and homes. In retaliation, the Indians would steal supplies from the pioneers crossing the area. Wild horses were also seen.

The grassland area near Morrill which later became Sun Springs Resort was prime buffalo territory. The ground in this area was rich, black soil with excellent pasture and hay for animals, plenty of water, and timber for building.

Grasshopper Invasion on September 8, 1874

The city was named after Edmund N. Morrill, a longtime friend of Thorton J. Elliot. Major E.N. Morrill served in the Civil War and later became a member of the U.S. House of Representatives for 4 terms, and then on to become Governor of Kansas (1894). He came from Portland, Maine in 1857. He was instrumental in organizing the Morrill and Janes Bank in Hiawatha. His wife was Elizabeth Brettum from Maine. She died during the Civil War, and he married Caroline Nash of Boston. In 1884 Morrill advertised he had 50,000 acres of ground for sale in the Hamlin area for $25 an acre. (See story of E.N. Morrill)

The closest connection to a railroad was at White Cloud, Kansas. Many rode that distance and walked the rest of the way to Morrill. (In 1870 the first railroad was completed.) In 1877, St. Joseph and Western Railroad erected a depot in Morrill and in 1878, The Grand Island Railroad, or the Union Pacific Railroad soon connected with the railroad for freight and passenger use.

In 1877, the company erected a railroad depot (now located at Sycamore Springs) and all future trains stopped in Morrill. T.J. Elliot was the first station agent and Miss Nettie Howe was the telegraph operator.

In 1952, the last passenger train ran through Morrill.

The main street was named Roxanna Street after the first wife of T.J. Elliot. At one time, this street was part of the road that led from White Cloud to Albany (north of Sabetha).

Morrill Main Street Parade before 1920 (See only thirty-nine stars on the flag.)

In 1913, a drugstore was owned by Ulysses. S. Davis, a druggist, and his wife Emma Jane (Gibbs). He had to retire from the post office because of a change in political parties. He developed an elixir called Davis's oil. It was a cure-all that was used by many. He had purchased the formula from Sam Haldeman. Ingredients included were oil red thyme, oil eucalyptus, oil sassafras, oil cajeput, oil cassia, and methyl salicylate, and

150 | Bonnie Dornes Hanni

This picture of the depot was taken July 31, 1952, when the last passenger train left the depot.

Morrill, KS, Train Station

sold for 90 cents a bottle. There weren't antibiotics in those days so many sought this to take care of illnesses. Melva Davis kept the "recipe" and the Greene's Drug Stores in Sabetha, Hiawatha, and Falls City purchased it from her.

The Church of the Brethren denomination had been organized in Schwarzenau, Germany in 1708 with migration to America beginning in 1719. They first settled in Germantown, Pennsylvania and organized a church in 1723.

In 1881, in Ohio, the German Baptist Brethren Church was divided into three divisions which became: The Old German Baptist Brethren (ultra-conservative); German Baptist Brethren (conservative); and the Brethren (liberal or progressive). It took several years for the division to ultimately occur. The division was caused by a disagreement about members attending high schools and colleges, Sunday School, revival meetings, and a salaried ministry. These extreme measures never were found in the Brown County churches.

In 1882 the congregation was divided into 3 parts: Pony Creek, Morrill, and Sabetha (north of Sabetha on old Highway 75).

The first church in the Morrill area was the Church of the Brethren, first known as Pony Creek. They held their meetings in the United Brethren Church near Sycamore Springs in the 1870s.

Morrill, KS, Parade

Morrill Schools

In the 1800s several schools existed in the area: #10 Township School; #16 Old Fairview; #54 Pennsylvania Avenue (northwest); Mulberry School (southwest); #57 Eagle School (at Sun Springs); #64 Spring Grove; #77 Lichty School; #20 at Sycamore Springs; and #52 Franklin School nearer to Sabetha. Number 34 was chosen as the school district, and the teachers had to present certificates of qualification and morality from a state inspector. (A competing school was formed as a private academy in 1882. It lasted 6 years but closed due to a lack of students. The building was later called the Old Morrill College.)

After the organization of the school (1857) taxes were required to support the school.

In 1878, it was decided to expand free education in Morrill and to combine a grade school and high school. In 1926 a gymnasium and more classrooms were added to the brick building and lunches started to be served in 1942. The building remained until 1910. A new brick building was built and remained until 1974.

The Morrill School Bell still stands on the grounds. Commencement exercises began to be held starting in 1908 in the Opera House located on the second floor of the Morrill State Bank.

In the school's history of 74 years, 689 students graduated from high school.

Morrill Grade & High School

In 1966 "unification" was the buzzword. Morrill school district closed and the children attended Sabetha school in 1969 and the school building was bull-dozed in 1974.

Far across the golden prairie, in her orange and black, stands the dear old Morrill High school, turning never back.

Push your colors ever forward, the orange and the black. Hail to thee, our Alma Mater, hail to M H S.

Some further interesting statistics:

- Pennsylvania Avenue families who came to Kansas in 1880: total children: 80
- Country Schools: District #54 and District #75 Schools
- Heikes School was located on Pennsylvania Avenue. Bertha Heikes Manche taught there as well as in the Sycamore country school.
- Morrill school's (1880) first teacher was Miss Kate Herbert (sister of Ewing Herbert from Hiawatha). Most of us remember Miss Melva Davis as the first grade teacher for many years.
- 8th Grade Graduation was held in Hiawatha for all Brown County graduates. Each student received a diploma from the County superintendent.

Entertainment

Cheerleaders: Gay Strahm, Bonnie Dornes, Sue McKim, 1966

In the late 1940s and early 1950s, benches were placed on Friday night to watch the free movie of the week during the summer. It was projected on a white square painted on the west side of a building. People would get there early to park their cars so they could participate. The movie was paid for by the city of Morrill. They paid a "traveling company" that also showed the same movie in Fairview and Hamlin and other small towns.

I remember going to Morrill with my folks to watch the movie on Friday evening. It was important that you got there early to park your car so you could watch the movie from the hood of your car.

All of the businesses were open in the evening. It was a time to do your grocery shopping at Myers Grocery Store. I remember penny gum called "Chum Gum." It had 3 pieces of gum for a penny.

This was also a time of community. The large white painted square stayed on the wall years after they stopped showing movies. A great time to see friends and family.

Bonnie's Cheerleader Letters

View of Hotel

Entrance to Hotel

Chapter 15
1984 - 2017

Bridge View of House & Hotel

Thousand Adventures, Inc.

Thousand Adventures, Inc. purchased Sycamore Springs in 1984 and began a security park selling memberships. Membership shares sold for $5000. Each share gave the owner entitlement to the 1250th acre of land.

The hotel was reopened in 1986 after extensive remodeling as a Bed and Breakfast facility. Much of the original structure and antique furnishings were preserved to delight the members who visited there.

The bedrooms were decorated with antique furniture and hand-made quilts, giving a feeling of nostalgia with memories of bygone days for each bedroom. Mary Dornes, who came back to work at Sycamore Springs, designed and made the curtains.

Mary Dornes sewed the curtains and arranged for the making of the quilts for the hotel rooms. Charles and Mary Dornes,

Hotel Sun Porch, 2nd Floor

Hotel Guest Room

Hotel Hallway, 2nd Floor

former owners of Sycamore Springs, continued to work for Thousand Adventure Co. for several years in the skating and concession stand area.

My parents liked being back working at the Springs and adding to the enjoyment of the people coming to visit. Old friends and people who came when they owned the Springs were a welcome sign to Mom and Dad.

Park homes were moved in to accommodate families not owning campers and several hookups were added for those fortunate enough to own a camper.

Thousand Adventure Company went into bankruptcy and Terry Tietjens became the owner once again until Betty and Dale Aue purchased Sycamore Springs in 1990.

Charles Dornes with Skates

Dale & Betty Aue, Owners

Dale and Betty Aue, farming neighbors of Sycamore Springs purchased the resort in 1990. They had a special bond with the facility. The family lived across the section and came to the Springs for many family activities.

The Aues came in with renewed vitality to make Sabetha an area destination attraction. The resort's roller rink, a large indoor facility attracted skaters of all ages.

Skating Rink Entrance

The roller rink still has the original oak wood flooring but a music system for skaters is now computerized with computerized music to flashing lights

Skate Couple from Marysville, Kansas

Skating Rink Floor

Roller Skates with Pom-Poms

View of Skating Rink

The swimming pool was not in operation at this time of their purchase. Federal regulations/codes required an extensive remodel. The cost was not feasible to remodel the pool. However, miniature golf was revitalized with refurbished obstacles. Tennis, volleyball, and basketball courts were included, also a baseball diamond and horseshoe pits.

Under the new ownership, other activities such as the Twisters Car Club of Sabetha came to Sycamore Springs for their annual June car show.

Betty Aue by Hotel

Dale cleared a hiking trail behind the house and up into the hills as an added attraction around the two natural springs.

60,000 gallons of water come from the springs every day

With the Church nearby, this facility could handle up to 350 people. The hotel was also used for weddings, and the clubhouse was used for group gatherings and wedding receptions. Over 150 weddings were performed in the Rock Creek Chapel. Church services were held on Sundays by churches within a 40-mile radius.

Dale and Betty Aue closed the resort following the 2017 season.

Betty Aue by Skating Rink

Spring #2 Water

Hiking Trail Entrance

Chapter 16
New Generation

Hotel & House Repainted

Kent Grimm Family, Owners Sycamore Springs 2020

Sycamore Springs was sold to Kent Grimm and his wife Molly, and Collin Grimm of Morrill, in 2020. Sycamore Springs's name was changed to Sycamore Springs Whitetail Ranch and the House and Hotel were renovated into an Airbnb.

The Grimm family had been searching for a place to raise whitetail deer, and after considering several other properties and options, the Grimms approached the owners Dale and Betty Aue.

Betty and Dale said, "We have always loved Sycamore Springs, but it looked like too big of a mountain to climb to re-open it." The Aue family children were getting older, and they were looking for ways for them to return to the family farm operation, and this seemed like a possible fit.

The Grimms raised game birds and raising whitetail deer was a viable addition to their operation. The whitetail deer later arrived at Sycamore Springs where the public can now view them. The mothers stay close to their babies.

With the new ownership, changes were coming to the facilities:

- The skating rink is remodeled and restored with open beams but remains the largest in Kansas.
- A kitchen has been added and the snack room doubled in size.
- Camping hookups are nestled in the trees. The RV Campground restrooms were remodeled
- with all new fixtures and flooring.
- The Rock Creek Chapel was moved to Albany. Since the resort has changed, it wasn't a building that fits in with the overall goal the Grimms are achieving.
- The Hotel was completely remodeled with the addition of a kitchen, new flooring, and central air conditioning. It sleeps 25 guests and will be rented out as a whole unit for families or large groups.
- Dornes (Charles Dornes) Hall Barn has been renovated for receptions and family gatherings. A kitchen was added to the barn.

Finally, the pool was removed due to too many regulations to bring it up to code. A splash park has been installed. No need for a lifeguard. Many features such as a slide were added.

The Grimm's goal for the resort is to offer a place for people to be able to bring their families and connect with their children's and grandchildren's hearts while making memories that will last a lifetime. It brings tranquility to the soul by hiking in the woods.

"Being in the great outdoors has a healing quality that is overlooked by the busy rat race of the time we live in," Kent Grimm expressed. Kent went on to say, "we need to slow down and enjoy the simple things that lift our hearts and spirits and put smiles on our faces."

They are a young family with young spirits. I am so pleased for them in all that they have done so far with Sycamore Springs. The word about Sycamore Springs has spread and people are coming, not only from the immediate area, but coming from a distance.

They are saving a piece of history for the next generation.

Sycamore Springs Memoir Bonnie Dornes Hanni 2023

There are two things I remember about Sycamore Springs. The first is that Sycamore Springs was my home for 17 years, 1949—1966. My mother and father built a successful business and included Dan and me as a working part of it. We grew up with restrictions like every other family. But we did have privileges that others don't experience by living at a resort. I had a good life.

The second thing I remember is what happened during these years set me on my course of life with all the experiences and the people that I meant along the way.

Not everyone can understand the dynamics of living at a historical resort. The rich, deep history came before we lived there. My life was lived in front of many people who came to Sycamore. Whether it was for sport or relaxation, in some way I experienced the same every day.

Sycamore Springs was a place of destination. Everyone who came had made plans to be there. My family could not decide who would come to visit us. Whether it was a last-day school picnic, family reunions, camp kids, Vacation Bible School camp, Bluebird groups, High School coach retreats, Church groups, or hay-rack rides, we served them to fulfill their needs. We met many wonderful people over the years and value their friendship. I never had to look for a boyfriend. All of the good-looking guys came to skate or swim.

There was never a lack of good (boy) friends.

We had many young people come that were from troubled families. My parents took them in and made them a

Whitetail Sign

part of our family. These kids would work for their room and board and live with us. The boys thought they became "hotshots," not knowing they were seeing happiness come back into their lives. There were always many people around us. As mentioned in my memoir, when we had a homesick camp kid, I became the little sister to bring comfort to that child.

Baby Deer

My family added only a small amount of history to the overall story of Sycamore Springs. To walk in the footsteps of those who came before is an honor. Dan and I explored every inch of this place; it was our playground.

I attribute my good health to drinking natural spring water for so many years. It has an acquired taste, high in sulfur. When we were at home, it was, "whose turn is it to run to the spring to fill the jug?" that we kept in the refrigerator. Summer or winter, with snow on the ground in a blizzard, someone did it. I believe in all the benefits of the crisp cool water that Alice Gray felt in her time.

In fact, I am sure Alice Gray and her friend Kickapoo Chief Chawkeekee would be amazed at the change in the "place of healing waters" known as Sycamore Springs.

***

There are many more historical facts that could be told. I apologize for any mistakes and the faint newspaper photos included. We don't know what lies ahead but each person connected with Sycamore Springs saw a future and had a vision of connecting to the past; we live there day after day until it is time to pass the dream on to the next dreamer.

Sycamore Springs has survived for the next generation. We press forward keeping the love of this place ready for those who come here for the first time or for those to keep the memory alive that so many experienced from days gone by.

The Gatekeepers of Sycamore Springs | 165

We can all say "I remember when…" That is what counts when you think of Sycamore Springs, and this has been a good experience to share my family life with you. I hope it triggers your memories and times at Sycamore Springs. It was a wonderful life living there and I want to thank everyone who came to visit our family. We hope you had a good time and that it was a wonderful and cherished memorable experience.

Steps by Hotel

House & Hotel at Night

Hotel Entrance

Kent Grimm Family,
Kent, Molly Kyle, Micha
Mylie, Addie, Selena, Aiden, Collin

Splash Park

Splash Park

Bonnie & Dan Dornes

Bonnie Dornes Hanni

References

AlbanyDays.org, Pinterest, Threshing Bee, Albany KS, Thrasher Tractor, Albany Historical Society, Albany Historical Museum.

Albany Historical Society, Albany Historical Museum, Albany School, Sabetha KS, 1965, photoshop.

Albany Historical Society, Albany Historical Museum, Albany Stores, Sabetha KS, 1965.

Albany Historical Society, Albany Historical Museum, Elihu Whittenhall, Sabetha KS, 1965.

Albany Historical Society, Albany Historical Museum, William Slosson, Sabetha KS, 1965.

Armitage, Katie H. Lawrence: Survivors of Quantrill's Raid, Images of America, 2010, Arcadia Publishing, Charleston, S.C. ISBN 978-0-7385-7799-9 Through a major grant from the Institute of Museum and Library Services, the Kansas State Historical Society, and the Kansas Collection of the University of Kansas present Territorial Kansas Online, a vivid illustration the time-period 1854 through January 1861, when Kansas entered the Union. National Debate About Kansas; Online 1854-1861; Territorial Kansas, A Virtual Repository For Territorial Kansas History

The Atchison Champion Newspaper, May 19, 1900, The Sycamore Mineral Springs-A Health Resort is fast gaining in public favor.

Betty Aue-Sabetha Herald-Sycamore Springs photos

BlackPast, B. 2022, 1854 Kansas Nebraska Act, BlackPast.org. https://www.blackpast.org/african-american-history/primary-documents-african-american-history/1854-kansas-nebraska act.

Boston Public Library, American Indian Village, Library of Congress, Public Domain, Unsplash

Boston Public Library, Kiowa, Summer 1869, Library of Congress, Public Domain, Unsplash

Brandscomb, Charles: volume I of Kansas: a Cyclopedia of State History, embracing events, institutions, industries, counties, cities, towns, prominent persons, etc. with a supplementary volume devoted to selected personal history and reminiscence. Standard Pub. Co. Chicago: 1912. 3 v. in 4: front., ill., ports.; 28 cm. Vols. I-II edited by Frank W. Blackmar.

Brougher, Shirley, Research for Sycamore Springs, Danny & Margaret Kendall, Shirley, Belinda

Brown Co. Genealogical Society reference. Greg Newlin, The Ancestor Trail, Sabetha Herald, Feb 2, 2005.

Brown County Recorder of Deeds, Hiawatha, Kansas

Common.wikimedia.org, Free-images.com/display/jameshenrylane.html, James H. Lane, Public Domain

References

Common.wikimedia.org, Free-images.com/display/john_wilkes_booth_portrait.html. John Wilkes Booth, jpg.Public Domain

Common.wikimedia.org, Free-images.com/display/American Indian village, 1879, Huffman Spotted Eagles. Public Domain

Common.wikimedia.org, Free-images.com/display/American Indians, pow_wow_nara-285618.jpg. Public Domain.

Common.wikimedia.org, Free-images.com/display/Pixabay, Public Domain, Abraham Lincoln man.jpg.

Common.wikimedia.org, Free-images.com/Lewis and Clark, jpg., Public Domain.

Common.wikimedia.org, Free-images.com/Merriweather Lewis & William Clark, jpg., Public Domain

Common.wikimedia.org, Free-images.com, Poster, Anti-Slavery Meeting, in Lawrence KS.jpg, Public Domain

Encyclopedia Britannica, https://www.Britannica.com/free-soilers

Encyclopedia Britannica, https://www.Britannica.com/Fox People

Encyclopedia Britannica.com Pawnee Scouts, Frank North, c. 1869. Library of Congress, https://www.britannica.com/topic/Pawnee-people#/media/1/447420/92497

Edward Sheriff Curtis: Visions of a Vanishing Race, Text by Florence Curtis Graybill and Victor Boesen, Promontory Press, NY., 1994.

Encyclopedia.com, James Henry Lane

Fairview Enterprise Press, Fairview, Kansas "You don't have to go to D.C. to drive down Pennsylvania Avenue", Interview of Langdon Livengood, Vol. 125, Feb 21, 2014.

Free-Images, Common Wikimedia.org, American Indian Village, 1879

Kent Grimm, Sycamore Springs Whitetail Ranch, 2022

Hesketh, Shawn, blog, Rock Cairn/Chimney Rock, photo, 2023

Hiawatha World Community Spotlight, June 24, 2014, The Dust Bowl "ill wind that blew no good."

History of Nemaha County, Kansas by Ralph Tennal, Sr, 1916, Standard Publishing Co., Lawrence, KS (Not in Copyright) Internet Archive. Ralph Tennal,Sr. 1872-1938.

www.history.com/topics/westward-expansion/westward-expansion

History.com/pilgrims-Mayflower-pilgrims aboard the Mayflower

Hughes, Patricia, Sabetha, KS, Photo, One of the springs, 1896, at Sycamore Springs, Sabetha Herald Memories.

www.Indiantraders.com,Native American Culture-Sweatlodges

John Brown to Bob Dole, Movers, and Shakers in Kansas History. Edited by Virgil W. Dean, 2006. University Press of Kansas, Lawrence, Kansas.

Andrew Johnson, What John Brown Did in Kansas, December 12, 1859

Kansas Country Living Magazine, Gliding Across Kansas, February 2011

Territorial Kansas, Kansas County Names, May 30, 1854.

State of Kansas, Territorial Kansas, May 30, 1854, Official State Seal

Kansas Historical Society. Governor biographies, E.N.Morrill, https://www.kansasmemory.org/item/200082#:Edmund Needham Morrill.

References | 171

Kansas Memory, Dog Chief, Pawnee Indian Scout, 1870-1889, #KSHS 210730, E99 P3.1DC

Kansas Historical Society, Kansas Herald of Freedom, 1854-1860, Topeka, Kansas

Kansas-Nebraska Act 1854, United States Statutes at Large: Treaties of the United States of America. Organization of Nebraska and Kansas Territories. Vol. 10, Page 277, Boston: Little Brown, 1855. The Avalon 1996.

Kansapedia Kansas Historical Society, Topeka, Kansas "Beecher Bibles", 2004.

Kansapedia, Kansas Historical Society, Seal of Kansas, 2020

Deborah Kmetz, U.S. General Land Office Surveyor's Field Notes

Barbara J. Kerr, Morrill: Facts, Fun, History, Class of 1954, Morrill High School, District #34 (Information taken from the 1911 Annals of Brown County, Kansas; Biographical Sketches of Morrill Township; The history of the state of Kansas; standard atlas of Brown County, Kansas)

Larry Gilbert-Sabetha Herald-Sycamore Springs photos

Lawrence Convention and Visitors Bureau- Kansas Humanities Council-Civil War on the Western Frontier, Lawrence, Kansas 2002

Legends of America/Indian Map/Kans¬¬as¬¬¬ https://legendsofkansas.com/wp-content/upoads/2020/05/KansasIndiansMap

Legends of America, Kansas Historical Society, Sioux Indians on Horseback, 1899

Legends of American, Frank W. Blackmar; Kansas: A Cyclopedia of State History, Standard Publishing, William Cutler, History of State of Kansas, James Lane Jayhawkers in the Civil War.

Legends of Kansas, Kansas Historical Society, Kansas Indian map

Learnodo-newtonic.com, Lewis & Clark Expedition, Louisiana Purchase Facts

Learnodo-newtonic.com, Thomas Jefferson & Napolean, Louisiana Purchase Facts

Library of Congress, Civil War photographs, 1861-1865 [Abraham Lincoln, three-quarter length portrait, seated and holding his spectacles and a pencil] Gardner, Alexander, 1821-1882, photographer,1865 Feb. 5. No known restrictions on publication.

Library of Congress, James Lane, 1855-1865, #LC-DIG-cwpbh-01175, No known restrictions on publication. Call #LC-BH82-4185

Library of Congress, John Brown, head and shoulders portrait, facing front/H. Sartorius, 1890-1910, No known restrictions on publication. #LOT 5910, no 88.

Library of Congress, Stephen A. Douglas, Senator from Illinois, 35th Congress, https://hdl.loc.gov/loc.pnp/ppmsca. 26791 No known restrictions on publication.

Library of Congress, John Wilkes Booth/ http://hdl.loc.gov/loc.pnp.print, Prints and Photographs.

Livingood, Langdon, Sabetha, Kansas, History, Research, Alice Williams Report 2014

Mayhew Cabin Museum, https://mayhewcabin.org, John Brown Cave

Clyde A. Milner II et al., The Oxford History of the American West (New

172 | References

York: Oxford University Press, 1994), 124-126.

Milwaukee Public Museum: www.mpm.edu, Kickapoo

Modern School Supply Co., Rowles, E.W.A., Chicago, Ill. 1919, The Missouri Compromise 1820, Historical geographical maps of the United States, Library of Congress.

National Underground Railroad Network to Freedom https://nps.org

Native American Culture-Sweatlodges-www.Indiantraders.com

Native Ministries International: https://data.nativemi.org/

Nebraska's Network to Freedom Sites, https://www.nps.org, John Brown

Network to Freedom KS-UGRR.pdf. Https://irp-cdn.multiscreensite.com, Lane Trail Map, Kansas map that marks the Underground Railroad Trail north to Sycamore Springs.

Greg Newlin-Sabetha Herald-Sycamore Springs photos

O'Bryan, Tony. "Bushwhackers" Civil War on the Western Border: The Missouri-Kansas Conflict, 1854-1865. The Kansas City Public Library. https://civilwaronthewesternborder.org/encyclopedia/bushwhackers.

Oregon Trail Marker in Sabetha Kansas, The St. Joseph News-Press, May 17, 1992

"In the Name of God, one is wrong", Leonard Pitts, Jr., Opinion, Lawrence Journal-World, Lawrence, KS, Nov 19, 2014. Smithsonian Magazine, Oct 2009, Day of Reckoning by Fergus M. Bordewich, pg. 62-69

The Reader's Companion to American History. Eric Foner and John A. Garraty, Editors. Copyright © 1991 by Houghton Mifflin Harcourt Publishing Company.

Sabetha Chamber.com, Be Wise Sign photo, Sabetha KS.

Sabetha History Committee. A History of Sabetha, Kansas, and Surrounding Area, 1854-1976. N.P.: 1976. (K/978.1/-N34/Sal3).

Smithsonian Magazine, Oct 2009, Day of Reckoning by Fergus M. Bordewich, pages 62-69

Smithsonian Magazine, History Channel

Stephen Douglas, The Reader's Companion to American History. Eric Foner and John A. Garraty, Editors. Copyright © 1991 by Houghton Mifflin Harcourt Publishing Company.

Strobridge Lith..Co., Cincinnati. Abraham Lincoln with Flags, LC-USZC4-1526, Partial print, photoshop enhanced. Emancipation Proclamation.

Sycamore Springs Sanitarium, SM Hibbard booklet Brochure, The Water Way to Health, Sycamore Mineral Springs Co., 1923

Teachushistory.org, Andrew Jackson and Indian Removal

teachingamericanhistory.org

Tennal, Ralph,Sr. History of Nemaha County, Kansas. Lawrence, KS: Standard Publishing Co., 1916. (K/978.1/-N34/T255). Narration/Interview by Alice Gray Williams, 2016. Alice, whom the Indians delight to call, "Soniskee," meaning- "Our Good Red Mother."

The Topeka Capital-Journal September 15, 2015, Jane Biles, Secret network in Kansas helped blacks escape slavery,1860.

References | 173

Tietjens, Jerry, History of Alice Gray report, History of Sycamore Springs
Tragic Prelude by John Steuart Curry illustrates John Brown and the clash of forces in Bleeding Kansas. (A mural painted by Kansan John Steuart Curry for the Kansas State Capitol building in Topeka, Kansas. It is located on the east side of the second-floor rotunda.)
Passion & Principle: John and Jessie Fremont, the couple whose power, Politics, and Love shaped nineteenth-century America, New York: Bloomsbury USA, 2007
Ushistory.org, Bloody Kansas, 23c. The Missouri Compromise, www.ushistory.org/us/31/asp
Morris W. Werner, Kansas Heritage, Lane's Trail, and the Underground Railway
Watson Museum, Lawrence KS, John Brown, reproduced by permission, photoshop/cropped, 2023.
Watson Museum, Lawrence KS, Ruins of Lawrence, Western Town Photo, reproduced by permission 2023.
Morris W. Werner, Northeast Kansas Trails
C. Albert White, A History of the Rectangular Survey System (Washington, D. C.: Government Printing Office, 1983), 9-16.
Wikipedia.org, Betsy Ross Flag
Wikipedia.org, State of Kansas, US flag

Newspaper Articles

Brown County Herald (Morrill), July 30, 1886, L.W. Phillipi, Visit Sycamore Springs
Brown County Herald (Morrill), August 6, 1886, Sycamore Springs
Brown County Herald (Morrill), August 6, 1886, Old Nick, Pony Creek
Brown County Herald (Morrill), August 13, 1886, Sycamore Springs
Falls City Journal, Falls City, Nebraska, Sycamore Springs great Indian Secret, March 5, 2002
Hiawatha World, Community Spotlight, Fun and History Abound at Sycamore Springs, June 21, 2005.
Lawrence Journal-World, Tim Rues and the small town with big history, Conrad Swanson, September 21, 2015.
The Sabetha Herald, August 7, 1886, Rush to Sycamore
The Sabetha Herald, June 14, 1888, Sycamore Springs (History)
The Sabetha Herald, July 11, 1889, The Fourth At Sycamore
The Sabetha Herald, August 16, 1889, The Flood At Sycamore Springs
The Sabetha Herald, Morrill News, The Sycamore Mineral Springs, July 20, 1900
The Sabetha Herald, February 24, 1916, Sycamore Springs Burns
The Sabetha Herald, May 29, 2002, Rock Creek Chapel re-opens
The Sabetha Herald Yesteryear photos, March 26, 2003; Oct 15, 2003; August 21, 2013; May 10, 2017; March 20, 2019.
The Sabetha Herald, February 2, 2005, The Ancestor Trail, Greg Newlin
The Sabetha Herald, Ancestor Trail, Part XIII: When Abraham Lincoln came

to the Kansas Territory, Greg Newlin, May 6, 2015; February 4, 2015, Part XII

The Sabetha Herald, May 27, 2015, Langdon & Elda Livingood photo

The Sabetha Herald, Memorial Day Supplement, May 20, 2015

The Sabetha Herald, Ancestor Trail, Part X: Continuation of Whitman Mission Route Entries, Greg Newlin, April 1, 2015, April 5, 2017

The Sabetha Herald, Ancestor Trail, Part XIII: Lost Diary Entries, Greg Newlin, April 6, 2016

The Sabetha Herald, Ancestor Trail, Part II: The Colorado Gold Rush Bust, Greg Newlin, May 4, 2016

The Sabetha Herald, Ancestor Trail, Part IX: John & Dorothy Bowlby Lanning Series, Greg Newlin, June 22, 2016

The Sabetha Herald, June 22, 2016, The Ancestor Trail, Part IX: John and Dorothy Bowlby Lanning Series, Greg Newlin, Norm Lanning & John Brougher photo

The Sabetha Herald, Sycamore Springs purchased, the name changed, March 24, 2021

The Sabetha Herald, Spring 1896, Sabetha Herald Memories

Topeka Capital-Journal, Connected Northeast Kansas, Local Business Sycamore Springs Resort, July 6, 2008.

http://www.kansastravel.org/albanyhistoricalmuseum.htm

https://www.americanhistorycentral.com/entries/kansas-nebraska-act-facts/ Stephen Douglas

https://www.thefamouspeople.com/profiles/abraham-lincoln-7.php

https://www.legendsofamerica.com/wp-content/uploads/2018/06/Kickapoo-Lodge.jpg

Kansas Memory, Dog Chief, Pawnee Indian Scout, 1870-1889, #KSHS 210730, E99 P3.1DC

https://legendsofkansas.com/wp-content/uploads/2020/05/KansasIndians-Map-700.jpg

https://www.legendsofamerica.com/wp-content/uploads/2018/11/1856PoliticalMap.jpg

Photo Citations

Page   Citation
Cover Kent Grimm Whitetail Ranch Upper Gate photograph 2022
Tribute Shirley Bozone photograph, Langdon and Elda Livengood
1   Ralph Tennel Sr. History of Nemaha County Kansas 1916-Alice Gray Williams
2   Bonita Hanni photograph Sycamore Springs Pony Creek
2   Bonita Hanni photograph Sycamore Springs, Sycamore Tree
2   Watson Museum, Lawrence KS, Ruins of Lawrence, Western Town Photo reproduced by permission. 2023
4   Bonita Hanni photograph, Sycamore Springs, Sycamore tree
5   Legends of America, Kickapoo Sweat Lodge
5   Bonita Hanni photograph Sycamore Springs, Blue Mud
6   Ralph Tennel Sr. History of Nemaha County Kansas 1916-George and Alice Williams
6   Ralph Tennel Sr. History of Nemaha County Kansas 1916-George Williams
9   History.com/pilgrims-Mayflower-pilgrims aboard the Mayflower
10   Wikipedia. org-Betsy Ross Flag
13   Kansas Indian Map, Legends of Kansas
16   Kansas Memory, Dog Chief, Pawnee Indian Scout
17   Sabetha Saint Anthony Hospital
19   Free-Images.com, Commons.Wikimedia.org, American Indian Village, 1879
20   Legends of America, Sioux Indians on horseback, 1899
21   Legends of Kansas, Kanza Men
22   Postcard-"Greetings", Bonita Hanni
24   Legends of Kansas, Indian Chiefs
24   Kickapoo Sculpture, Horton, KS, Bonita Hanni photo 2022
25   Buffalo, Kickapoo Nation Reservation, Horton, KS, Bonita Hanni photo 2022
29   Sod House Homestead, Ralph Tennel Sr, History of Nemaha County Kansas 1916
30   Wagon Train, Ralph Tennel Sr, History of Nemaha County Kansas 1916
32   Edmond Needham Morrill, Kansas Memory
34   Louisiana Purchase Facts, Learnodo-newtonic.com Lewis & Clark Expedition
34   The Missouri Compromise 1820. Library of Congress
35   Louisiana Purchase Facts, Learnodo-newtonic.com Thomas Jefferson & Napolean

176 | Photo Citations

36	Merriweather Lewis and William Clark.jpg Public Domain-US
37	Kansas-Nebraska Act, 1854, BlackPast.org.
38	Constitution Hall, Lecompton KS, Bonita Hanni photo, photoshop enhanced
38	First Territorial Capital, Lecompton KS, Bonita Hanni Photoshop enhanced
39	Poster calling for an Anti-Slavery Mass Meeting in Lawrence, Kansas. jpg, Public Domain
40	Stephen A. Douglas, Anthony, Edward, 1818-1888
41	Lincoln, Abraham, 1809-1865, Library of Congress print, cropped/photoshop
42	Kansaspedia, Kansas Historical Society, Seal of Kansas 2020
42	Abraham Lincoln and his Emancipation Proclamation with flags
43	Massachusetts Street, Lawrence, KS, Watson Museum, Lawrence, KS 2022 copy photo
44	Main Street on Fire, Lawrence, KS Photo taken at Watkins Museum, Lawrence KS Used by permission 2023
44	Bushwackers, photoshop enhanced
45	William Quantrill, Photo reproduced and photoshop enhanced, Watson Museum, Lawrence KS, 2023
46	James Lane Jayhawkers in the Civil War, Legends of America
47	Haines Clothing Store, Sabetha KS, Ralph Tennel Sr, History of Nemaha County Kansas, 1916
47	KU Flag Banner, Bonita Hanni personal item, photo
48	Tragic Prelude Mural by John Steuart Curry Illustrates John Brown
49	John Brown, Watson Museum, Lawrence KS
50	Mayhew Cabin, Nebraska City, NE John Brown Cave
50	Mayhew Cabin, Nebraska City, NE, John Brown Cave
51	Mayhew Cabin, Nebraska City, NE, John Brown Cave, Ladder to Escape
51	John Wilkes Booth, Library of Congress
53	Slavery Wanted Poster, 1860, Topeka Capital-Journal
54	Albany Stores, Albany Historical Museum, Sabetha KS
54	Albany School, Albany Historical Society, photoshop
55	William Slosson, Albany Historical Society, 1965, Sabetha KS
56	Elihu Whittenhall, Albany Historical Society, 1965, Sabetha KS
56	Nemaha County Grain Harvest, History of Nemaha County Kansas, Tennel
57	Grain Harvest, Albany KS, History of Nemaha County Kansas, Tennel
58	Threshing Bee, Albany KS, Thrasher Tractor, Pinterest
58	Threshing Bee, Albany KS, Thrasher Machine, Pinterest
58	Threshing Bee, Albany KS, Thrashing Days, Pinterest
59	Main Street 1930-1940, Sabetha, KS, History of Nemaha County Kansas, Tennel
59	Oregon Trail, St. Joe Road, St. Joeseph MO Newspaper, 5-17-1992
59	Sabetha Water Tower, Oregon Trail Street Sign, St. Joseph MO Newspaper, 5-17-1992

Photo Citations | 177

60	Be Wise Owl Sign, Sabetha, KS, Sabetha Chamber
60	Sabetha Murdock Hospital, Sabetha, KS
60	Mary Cotton Library, Sabetha KS, Library.org
61	Old Elementary School, Sabetha, KS, History of Nemaha County Kansas, Tennel
61	Ernie Block Movie Theater, Charles & Mary Dornes front row with Leona & Bob Murchison, Sabetha, KS, Bonita Hanni
62	A&W Rootbeer Mugs, Ebay
62	Sabetha High School, Sabetha, KS, History of Nemaha County Kansas, Tennel
63	James H. Lane, Common Wikipedia
64	Rock Cairn/Chimney Rock, Hesketh blog
65	Lane Trail Map, Kansas Map marking the Underground Trail north to Sycamore Springs, UGRR Network to Freedom
67	Sycamore Springs Land, Aerial View, Kent Grimm
68	State of Kansas, US Flag, Wikipedia.org
69-70	Owners Deeds for Sycamore Springs Land, 1851-2020, Langdon Livengood
71	Mr. & Mrs. Jacob Bougher, proprietors, Kent Grimm Whitetail Museum
77	One of the Springs, 1896, Patricia Hughes, Sabetha Herald Memories
78	Three Story Hotel, Sycamore Springs, Sabetha Herald
78	Water Tower Reservoir for Hotel, Bonita Hanni
79	Spring 1896, Sabetha Herald Memories
79	E.V. Kauffman, Proprietor of Sycamore Springs, Bonita Hanni
80	Sycamore Mineral Springs Brochure Ad, Bonita Hanni
80	Sycamore Mineral Springs Hotel Postcard, Bonita Hanni
81	Hotel Early 1900s Sabetha Herald Newspaper Photo, Submitted by Betty Aue
81	Sycamore Mineral Springs Hotel Side Entrance, Sabetha Herald, Larry Gilbert
82	Hotel Burned 1916, Sabetha Herald Newspaper
82	Women by the Spring, Larry Gilbert photo, Sabetha Herald
82	Four Story Hotel, Whitetail Museum, Kent Grimm, 2022
83	Hotel with Bridge Postcard, Bonita Hanni
83	Hotel Burned 1916, Sabetha Newspaper
84	Guest on the Hotel Porch, 1912
84	Old Spring Postcard, Bonita Hanni
84	Pennyland, Sycamore Springs, Terry Tietjens photo
85	Carousel Merry-Go-Round, Terry Tietjens photo
85	Kansas Corporation Seal for Railroad, Sabetha History Committee, 1976
86	Post Office Replica, Bonita Hanni photo, 2018
86	Water Delivery Truck, The Water Way to Health Brochure-Sycamore Mineral Springs Co., 1923
87	Gathered around the Upper Spring No.2, The Water Way to Health Bro-

Photo Citations

	chure-Sycamore Mineral Springs Co., 1923
87	View of Hotel & Doctor Clinic, The Water Way to Health Brochure-Sycamore Mineral Springs Co., 1923
88	The Doctor's Clinic, The Water Way to Health Brochure-Sycamore Mineral Springs Co., 1923
88	Partially rebuilt Hotel, The Water Way to Health Brochure-Sycamore Mineral Springs Co., 1923
89	Dr, Samuel M. Hibbard, The Water Way to Health Brochure-Sycamore Mineral Springs Co., 1923
89	Clemons Rucker, Doctor, The Water Way to Health Brochure-Sycamore Mineral Springs Co., 1923
90	Roy Spring, Investor, The Water Way to Health Brochure-Sycamore Mineral Springs Co., 1923
90	John Koch, Investor, The Water Way to Health Brochure-Sycamore Mineral Springs Co., 1923
90	Owners by Spring, The Water Way to Health Brochure-Sycamore Mineral Springs Co., 1923
91	Sanitarium Newsletter Left Side, Bonita Hanni
91	Sanitarium Newsletter Right Side, Bonita Hanni
91	Putting Final Stucco on Buildings, 1930, Samuel Hibbard Newsletter
92	Sycamore Mineral Springs Postcard, Bonita Hanni
92	Exam Table, Bonita Hanni, 2018
92	Therapy Table, Bonita Hanni, 2018
93	Brochure, Water Way to Health, Samuel Hibbard, Bonita Hanni, 2018
93	Example of a Water Therapy Bath House, Hydrotherapy Facility Internet Search, Mental Health Treatment
93	Hotel Guest Room Health, The Water Way to Health Brochure-Sycamore Mineral Springs Co., 1923
93	Dining Room, The Water Way to Health Brochure-Sycamore Mineral Springs Co., 1923
94	Cabins for Patient Health, Bonita Hanni, 2018
94	Sycamore Springs Sanitarium Postcard, Bonita Hanni
95	Black Sunday in the Midwest, Hiawatha World Community Spotlight Dust Bowl 6-24-2004
95	Kansas Farm Dust Bowl, Hiawatha World Community Spotlight Dust Bowl 6-24-2004
95	Dust Bowl Masks, Hiawatha World Community Spotlight Dust Bowl 6-24-2004
96	Laurel Slayton Family during the Great Depression, Bonita Hanni Grandparents, Mary Slayton Dornes on Fender
97	Noah Edelman, Nila Edelman Snyder photo
98	Lillian, Nila, Morris Edelman, Nila Edelman Snyder photo
98	Camp House, Bonita Hanni photo 2018
98	Camp House with girls, Nila Edelman Snyder photo
99	Café House Entrance, Bonita Hanni photo
99	Café House Entrance with Nila and friends, Nila Edelman Snyder photo

Photo Citations | 179

100	Noah Edelman Horses, Nila Edelman Snyder photo
100	Noah Edelman cattle in Pony Creek, Nila Edelman Snyder photo
100	Empty Swimming Pool, Nila Edelman Snyder photo
101	Nila Edelman Snyder by the pool, Nila Edelman Snyder photo
101	Swimmers in the Pool, Sabetha Herald Yesteryear, Greg Newlin photo
102	The Spring, Bonita Hanni photo
102	Chuck Dornes Skate Boy, Bonita Hanni photo
102	The Skating Rink, Bonita Hanni photo
102	Clamp-on Skates, invaluable.com/auction/vintage-metal-roller-skates
103	Camping Brochure, Terry Tietjens, Bonita Hanni submitted
103	View of the House from the Bridge, Bonita Hanni photo
103	View of the Bridge, Bonita Hanni photo
104	Our Home, Bonita Hanni photo
105	Danny and Bonnie Dornes 1949, Bonita Hanni photo
105	Danny and Bonnie Dornes 1950s, Bonita Hanni photo
105	Danny and Bonnie 1955, Bonita Hanni photo
105	Charles and Mary Dornes at the Skating Rink, 1950s, Bonita Hanni photo
106	The Shop, 1959, Bonita Hanni photo
106	Dornes Hall, Bonita Hanni photo
107	Life Magazine, JFK Memorial 1963, Campaign Button, Bonita Hanni
108	Chuck Dornes, 1955, Bonita Hanni photo
109	Mary Dornes, 2019, Bonita Hanni photo
110	Debbie Reynolds Paper Dolls, Bonita Hanni
111	Bonnie Toy Collection, Bonita Hanni
111	Superman Comic Book, Comic Art Fans
112	Gideon New Testament, Bonita Hanni
112	Moody Bible Institute Banner, Bonita Hanni
113	Gazebo Spring, Winter, Bonita Hanni
113	Winter at the House, Bonita Hanni
113	Winter on the Front Steps, Bonita Hanni
114	Winter at the Skating Rink, Bonita Hanni
114	Winter at the Swimming Pool, Bonita Hanni
114	Winter on Pony Creek, Bonita Hanni
115	1966 Morrill High School Yearbook, Bonita Hanni
115	4-H Pin, Bonita Hanni
116	Danny and Margaret Kendall, Shirley Kay, Belinda 1950s, Shirley Brougher
116	Danny and Margaret Kendall, Shirley Kay, Belinda 1966, Shirley Brougher
116	Charles and Mary Dornes, Dan and Bonnie, 1966, Bonita Hanni
117	Aerial View of Swimming Pool postcard, Bonita Hanni
118	Ray Harris, Gayla Manche, Bonnie Dornes, Bonita Hanni
118	Dan Dornes, Bonita Hanni
118	Bonnie Dornes, 1956, Bonita Hanni
119	Dan, Mary, Bonnie 2018, Bonita Hanni

180 | Photo Citations

119	Swimming Pool ready for the Summer, Mary Dornes photo
120	Swimming Pool Towards the Deep End, Mary Dornes photo
120	1950s Skating Rink and Swimming Pool, Bonita Hanni
121	Swimming Show Program, 1961, Sabetha Herald Newspaper, Bonita Hanni
122	Mary Dornes in the Rack Room, Bonita Hanni
122	Bonnie Dornes by the Pool, Bonita Hanni
123	Bonnie Dornes Life-Saving Badges, Bonita Hanni
124	Kids by the Pool, Sabetha Herald, Yesteryear, 8-21-2013, Greg Newlin
125	Flood by the Shop, Dan Dornes
125	Flood by the House, Dan Dornes
125	Pony Creek Flood, Dan Dornes
126	Kids in Pickup, Bonnie Dornes in kerchief, Bonita Hanni
126	Paper mâché Pumpkin, liveauctioneers.com/vintage Halloween papier mâché jack-o-lantern
126	Coca Cola Sign, Etsy photo
127	Coke Water Cooler 1940s, Collector's Weekly
127	1955 Record Player, Oldest.org
128	Chubby Checkers, The Twist Record Album, Stereogum
129	Bonnie and Danny with Badminton, Bonita Hanni
129	Danny Dornes, First Day of School, Bonita Hanni
129	Danny Dornes, Age 9, Bonita Hanni
130	Vauxhall Car, gmauthority.com/cars of the Vauxhall heritage collection
133	Camp Lanyards, Bonita Hanni
133	Mangle Iron, Pinterest/vintage mangle iron machine, The Good Ole Days
134	Terry & Jerry Tietjens, Terry Tietjens photo
135	Camp Store and Antique Store, Terry Tietjens photo
135	Skating Rink with Disco Lights, Terry Tietjens photo
135	Spring Gazebo, Terry Tietjens built, Bonita Hanni
135	Spring Gazebo near the Hill, Bonita Hanni
135	Water Chemical House in Winter, Bonita Hanni
136	Camper Area, Thousand Adventures, Inc. brochure. Whitetail Ranch, Kent Grimm
136	Promotional Brochure, Terry Tietjens
136	Sycamore Springs, Square Dance, Terry Tietjens
137	Morrill Train Depot, Bonita Hanni
137	Rock Creek Church, Bonita Hanni
137	Church Organ, Bonita Hanni
138	Kansas State Department of Health, 1970, Water properties, Terry Tietjens
139	Rock Creek Church, Bonita Hanni
139	Farmer Community, Sabetha Herald Newspaper, Memories, 3-19-2003, Langdon Livengood
141	Yellow Cross made by Jimmy Kesler, Bonita Hanni
142	Sommerset Pennsylvania Map, City-Data.com

Photo Citations | 181

143 US Commemorative Stamp, Susan Steiner-German Information Center, Richard Schlecht artist, 2014
144 Livengood Homestead, Langdon Livengood
145 Chautauqua Speaker, Chautauqua Institution/history, Library of Congress Public Domain
146 Bertha Heikes, Teacher, Sabetha Herald Newspaper, Birthday Celebration
146 1938-1939 School Kids, Pleasant Valley School District 10, 2014, Langdon Livengood
146 1938-1939 School Kids, Student Names, Pleasant Valley School District 10, 2014, Langdon Livengood
147 1939-1940 School Kids, Pleasant Valley School District 10, 2014, Langdon Livengood
147 1939-1940 School Kids, Student Names, Pleasant Valley School District 10, 2014, Langdon Livengood
147 Bertha Heikes Manche, Bonita Hanni
147 Gayla Manche, Bonita Hanni
150 Morrill Train Depot, Barbara Kerr, Morrill High School, Morrill Facts, Fun, History
150 Morrill KS Parade, Langdon Livengood research, Sabetha Herald, Yesteryear 8-7-2002
151 Morrill Grade and High School, Bonita Hanni, 1966 Yearbook
152 1966 Cheerleaders, Gay Strahm, Bonnie Dornes, Sue McKim, Yearbook, Bonita Hanni
153 Bonnie's Cheerleader Letters, 1966-1967, Bonita Hanni
154 View of Hotel, Bonita Hanni 2018
154 Entrance to the Hotel, Bonita Hanni 2018
155 Bridge View of House and Hotel, Bonita Hanni 2018
156 Hotel Sun Porch, 2nd Floor, Bonita Hanni 2018
156 Hotel Guest Room, Bonita Hanni 2018
156 Hotel 2nd Floor Hallway, Bonita Hanni 2018
157 Charles Dornes with Skates, Worked for Thousand Adventures, Inc., Bonita Hanni
157 Skating Rink Entrance, Thousand Adventures, Inc., Bonita Hanni
158 Skating Couple from Marysville KS, Kansas Country Living, Gliding Across Kansas, February 2011, Betty Aue
158 Skating Rink Floor, Thousand Adventures Inc., Bonita Hanni
158 Roller Skates with Pom-Poms, smithauctions.com/j higgins 1950s Roller Skates with pom-poms
158 View of Skating Rink, Bonita Hanni
159 Betty Aue by Hotel, Hiawatha World Newspaper Community Spotlight 6-21-2005
159 Betty Aue by Skating Rink, Hiawatha World Newspaper Community Spotlight 6-21-2005
160 Number 2 Spring Runoff, Hiawatha World Newspaper Community Spotlight 6-21-2005

182 | Photo Citations

160 Hiking Trail, Developed by Dale and Betty Aue, Bonita Hanni photo
161 Hotel and House Repainted, Kent Grimm, Whitetail Ranch
163 Whitetail Entrance Sign to Sycamore Springs, Kent Grimm
164 Baby Deer, Whitetail Ranch, Kent Grimm
165 Steps by the Hotel, Bonita Hanni
165 House and Hotel with Night Lights, Whitetail Ranch, Kent Grimm
165 Hotel Entrance, Whitetail Ranch, Kent Grimm
166 Kent Grimm Family, Sabetha Herald Newspaper, 3-24-2021
166 Splash Park, Whitetail Ranch, Kent Grimm
166 Splash Park Mural, Whitetail Ranch, Kent Grimm
167 Dan Dornes and Bonita Hanni, 2023
167 Bonnie Dornes Hanni, 2023

Author Biography

Historical research is at the center of Bonnie Dornes Hanni's life. *The Gatekeepers of Sycamore Springs, Kansas Historical Journey to Healing Water*, culminates the lives of the owners of the public historical resort and how each generation made improvements for the times while preserving the nature of the crystal-clear spring water.

- One woman's story, Alice Gray Williams, and the friendship with the Kickapoo Indians and their secret "healing water."
- History of the American Indians who used the natural spring water for healing.
- Pioneers traveling through for fresh water on their way to discover gold in California.
- John Brown, an abolitionist, used Underground Railway in Nemaha County.
- James Lane Freedom Trail helped slaves travel through Kansas to achieve freedom.
- Owners of the Public Resort 1880-2022.
- Sycamore Springs Sanitarium and Hotel for water therapy.
- Recreational area with Skating Rink and Swimming Pool, Splash Park.
- Campgrounds and Event Center.
- Historical Museum in Train Depot.

These are a few of the highlights that took place at this sycamore land. There is a deep rich history of Sycamore Springs.

Bonnie has also written and illustrated a children's book, *Where is Bob?*, a story relating our currency to historical places in the United States, found on Amazon. She has several other books in the making.

Bonnie lives in Lawrence, Kansas ~~and is presently the House Director for Tri Delta Sorority at the University of Kansas.~~

Printed in the USA
CPSIA information can be obtained
at www.ICGtesting.com
JSHW070947201223
54067JS00019B/144